OECD STEEL OUTLOOK
1999/2000

1999 *Edition*

ORGANISATION FOR ECONOMIC CO-OPERATION AND DEVELOPMENT

ORGANISATION FOR ECONOMIC CO-OPERATION AND DEVELOPMENT

Pursuant to Article 1 of the Convention signed in Paris on 14th December 1960, and which came into force on 30th September 1961, the Organisation for Economic Co-operation and Development (OECD) shall promote policies designed:

- to achieve the highest sustainable economic growth and employment and a rising standard of living in Member countries, while maintaining financial stability, and thus to contribute to the development of the world economy;
- to contribute to sound economic expansion in Member as well as non-member countries in the process of economic development; and
- to contribute to the expansion of world trade on a multilateral, non-discriminatory basis in accordance with international obligations.

The original Member countries of the OECD are Austria, Belgium, Canada, Denmark, France, Germany, Greece, Iceland, Ireland, Italy, Luxembourg, the Netherlands, Norway, Portugal, Spain, Sweden, Switzerland, Turkey, the United Kingdom and the United States. The following countries became Members subsequently through accession at the dates indicated hereafter: Japan (28th April 1964), Finland (28th January 1969), Australia (7th June 1971), New Zealand (29th May 1973), Mexico (18th May 1994), the Czech Republic (21st December 1995), Hungary (7th May 1996), Poland (22nd November 1996) and Korea (12th December 1996). The Commission of the European Communities takes part in the work of the OECD (Article 13 of the OECD Convention).

Publié en français sous le titre :

PERSPECTIVES DE L'ACIER DE L'OCDE 1999/2000
Édition 1999

FOREWORD

The OECD Steel Committee decided to undertake studies of the steel market and its outlook at its first meeting in 1978. Since that date, a report has been published yearly, beginning with *The Steel Market in 1978 and the Outlook for 1979*.

This report was prepared by Mr. Franco Mannato of the OECD Secretariat. The Steel Committee examined the report, which is based on data received before 31 March 1999. It is published on the responsibility of the Secretary-General of the OECD.

TABLE OF CONTENTS

INTRODUCTION

At its 52nd meeting, in the spring of 1998, the Steel Committee decided, when adopting its programme of work, that a report on steel market trends in 1998 and the outlook for 1999 and 2000 would be drawn up in early 1999. After being discussed by the Steel Committee, the report will be published on the responsibility of the Secretary General of the OECD, as in previous years.

Certain delegations to the Steel Committee provided statistics and other information on market developments in their countries, and the Secretariat has taken these into account. Since, however, it had to produce a coherent world outlook, it is possible that the text and the estimates may differ somewhat from those provided by the various delegations. It is therefore the Secretariat that is responsible for the forecasts.

Following the admission of the Czech Republic, Hungary, Korea and Poland as Member countries of the OECD during 1996, the statistics for these countries have been included in the OECD total and, as far as possible, in order to maintain a degree of coherence, the historical data have been recalculated on that basis. Also, since Brazil became a full participant in the Steel Committee in 1996, statistics for that country have been added to most of the tables and removed from those for the Latin American zone.

As a result of these changes, the data for the Czech Republic, Hungary and Poland have been removed from the Central and Eastern Europe zone and have been included in the "other Europe" zone. As regards the European Union, only the UE(15) zone remains and the historical data have been recalculated as far as possible.

As a result of the financial crisis in Asia, the Secretariat has split the "other Asia" zone into two parts: the "ASEAN(5)" area which covers Indonesia, Malaysia, the Philippines, Singapore and Thailand, and the "rest of Asia", including North Korea, which is no longer grouped with China.

As far as possible, in addition to data for the New Independent States (the former Soviet Union) overall, the Secretariat has also tried to provide a breakdown for Russia, The Ukraine and the other NIS.

The report has been drawn up using the information received and the statistics available as at 31 mars 1999, and includes updates provided by Member countries before 18 May 1999.

NOTES ON THE MAIN FEATURES OF THE STEEL MARKET IN 1998, 1999 AND 2000

The main quantitative results for the steel market in 1998 and probable trends in 1999 and 2000 are contained in the statistical annex to the present document. The main developments in the market may be summarised as follows:

1998

Apparent steel consumption

- World: World steel consumption, which had risen sharply in 1997 (+ 6.8%), fell by 2.3% in 1998. This decline reflects the combined effects of the financial crisis in Asia, which caused a steep fall in steel consumption in that area from the second half of 1997 onwards, and the crisis that affected Russia and other NIS during the summer of 1998. Uncertainty over the possibility that the crisis might spread to Latin America also had a negative impact on world steel demand.

- OECD: After the significant rise in steel consumption of more than 8% in 1997, demand fell by 0.6% in 1998. This slight decline mainly reflected the severe fall in steel consumption in the Asian Member countries, Korea and Japan, which reported falls in consumption of around 17.2% et 37.1% respectively, while steel consumption in North America and Europe continued to rise significantly. In 1998, with a total of 413.1 million tonnes in finished product equivalent, apparent steel consumption in the OECD area remained at the second highest level achieved hitherto.

- While consumption fell in Korea and Japan, it remained at its 1997 level in Oceania. It rose by 8.6% in the United States, by more than 9% in Canada, and by as much as 22% in Mexico. Steel consumption was up by 8.7% in the European Union and by just over 7% in other European countries.

- In Brazil, 1998 saw a further increase in apparent steel consumption of around 3.3%, following the strong 19% rise reported in 1997. However, as a result of the crisis in Russia over the summer, the situation deteriorated towards the end of the year.

Graph 1. **World apparent steel consumption**
Million tonnes of product equivalent

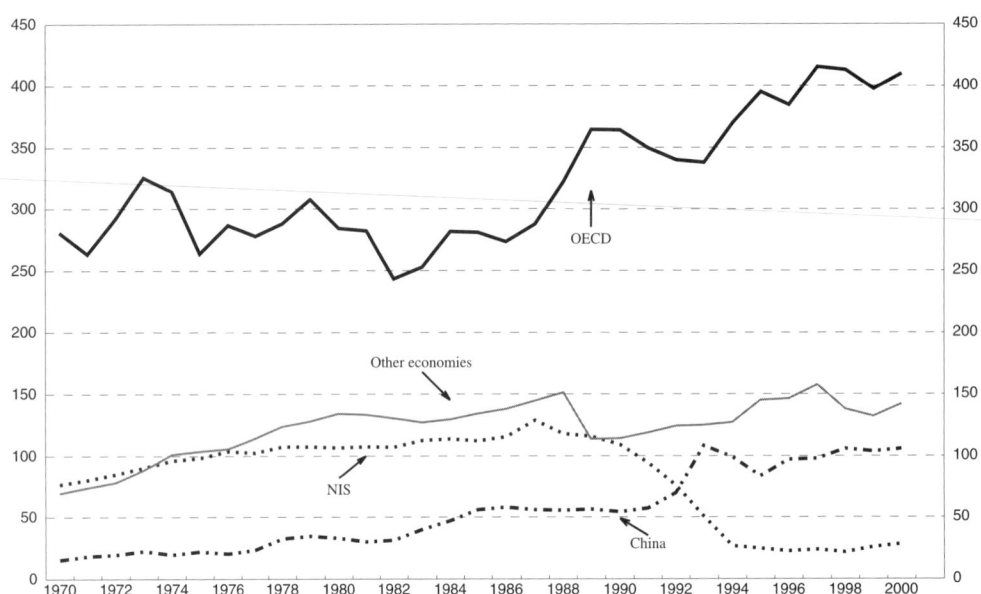

Source: OECD.

– As for other areas, in the other Latin American countries demand for steel in 1998 was down by 7.1% on 1997. In South Africa the fall was in the region of 14.6%, whereas the rest of Africa reported a fall in steel consumption of around 2%. In the Middle East, steel consumption was down by 12%, the first fall to be reported in the region in ten years. Similarly, steel consumption in India fell by 2.7%. As for the five countries grouped under the heading ASEAN(5), the effects of the financial crisis they experienced in the course of 1997 caused steel consumption to drop by rather more than 36.8%, nearly 12 million tonnes down on 1997 and 15 million tonnes down on 1996. In the other Asian countries, apparent steel consumption was down by 5.2%, with the fall in Chinese Taipei remaining fairly limited.

– The downward trend in steel consumption finally reversed itself in the New Independent States (ex-USSR) in the course of 1997, but the recovery failed to stand up to the crisis that broke in the summer of 1998. In the NIS as a whole, apparent steel consumption fell by 8.7% in 1998 to less than 22 million tonnes, the lowest level in its history. Consumption declined least in Russia, where it was down by no more than 4.5%; but it fell by 7% in The Ukraine and by more than 30% in the other NIS.

– In the non-OECD countries of Central and Eastern Europe, apparent steel consumption fell by 4.4% in 1998, but trends were very varied, with demand remaining stable in Romania, falling by 25.2% in the Slovak Republic and rising by 8.2% in Bulgaria.

– In China, demand for steel continued to increase steadily by around 8.4%, or 8.3 million tonnes more than in 1997.

Growth in apparent steel consumption (AC) and estimated growth in real consumption of steel (RC) and total steel stocks held by steel producers, consumers and merchants[1]

	OECD	Rest of OECD	Total for United States, EU and Japan				
					Total stocks of steel		
	AC	AC	AC	RC	Yearly change	End-year level	
	In million tonnes of finished product equivalent						In weeks of real consumption
1989	364.6	72.1	292.5	296.8	-4.3	87.6	15.3
1990	364.0	65.8	298.2	294.5	+3.7	91.3	16.1
1991	349.6	64.7	284.9	288.6	-3.7	87.6	15.8
1992	340.0	63.0	277.0	279.1	-2.1	85.5	15.9
1993	337.9	69.8	268.1	270.7	-2.6	82.9	15.9
1994	369.9	79.8	290.1	290.1	0	82.9	14.9
1995	394.7	87.7	309.7	304.1	+5.6	88.5	15.1
1996	384.6	86.2	298.4	304.7	-6.3	82.2	13.3
1997	415.6	94.4	321.2	320.2	+1.0	83.2	13.5
1998[e]	413.1	85.0	328.1	317.5	+10.6	93.8	15.4
1999[p]	397.5	88.2	309.3	314.0	-4.7	89.1	14.8
2000[p]	410.2	93.6	316.6	320.1	-3.5	85.6	13.9

e. Estimate.
p. Forecast.
1. In previous years, figures for apparent steel consumption can be derived, as they have in this report, from the data available on steel production and trade. Variations in apparent consumption are due to variation in real consumption and/or changes in the total steel inventories maintained by steel producers, consumers and merchants. Data regarding the level of, or annual variations in, both these parameters, however, are far from complete. The figures given for real consumption and annual variations in total stock levels should therefore be taken as "reasonable" estimates of two inter-related factors. Furthermore, in calculating the level of total steel stocks in tonnage terms by the end of 1984, it has been assumed that the stocks were equal to 18 weeks for estimated real consumption for that year (*i.e.* 8 weeks for producers and 10 weeks for consumers and merchants). For the years after 1984, the level of total steel stocks was first calculated in terms of tonnage, based on the estimated annual variation, and subsequently expressed in terms of weeks of real consumption.
2. As from 1995, data concerning the EU are related to EU(15) and the OECD total includes the new countries: the Czech Republic, Hungary, Korea and Poland.
Source: OECD.

- The fall in demand for steel products in the Asian area produced serious upheavals in the traditional trade flows in steel products. Significant quantities of steel were accordingly diverted towards more buoyant markets, though the consequence was a substantial fall in prices. As a result, overall steel stocks in the OECD area rose sharply in 1998, and this trend led to a fall in real steel consumption in the OECD area by around 3% in 1998.

Steel trade

- World trade in steel (excluding intra-EU trade) fell by 2.5% (by volume) in 1998 compared with 1997. World trade in steel accounted for 26.6% of world steel consumption in 1998.

- Steel exports from the OECD area remained at the 1997 level of 101.1 million tonnes, whereas imports rose by 13.8%, or 11.7 million tonnes more than in 1997. Consequently, net OECD steel exports were down by 74.5%.

- In the United States, the record demand for steel products produced an unprecedented rise in steel imports, which increased by 33% in 1998 to 38.3 million tonnes, or 9.4 million tonnes more than in 1997. Exports of steel products, however, fell by 11.3%. Despite a very healthy domestic market, the share of imports on the United States market rose from 26.7% in 1997 to 32.7% in 1998.

- In the European Union, despite the buoyancy of demand, the steel trade was seriously disturbed by disruption on the international scene. The balance of trade in steel products collapsed and the surplus of 11.7 million tonnes achieved in 1997 fell to 0.5 million tonnes in 1998, amounting to a net loss of 11.2 million tonnes, or 4 billion ECUs in one year.

- In Japan total steel imports fell by 30.7% in 1998 to 6.6 million tonnes, their lowest level since 1986. In the same period, steel exports rose by 17.6% to 27.7 million tonnes, their highest level since 1988.

- As for market economies outside the OECD area, net steel imports fell overall by nearly 27%, 16.3 million tonnes down on 1997. Net imports were down in Latin America and the Middle East, but the decline was greater in the Asian countries, falling by 45% in the ASEAN(5) and by 20% in the other Asian countries.

- Net imports to China increased by 51%, following a steep decline in exports of more than 65%, whereas imports fell by around 3.7%.

- Net exports from the NIS as a whole showed a slight decline of about 7%. While steel exports from Russia fell by 9.1% and those of The Ukraine by 4.2%, a significant proportion of these exports, initially intended for Asian markets, were diverted to the buoyant markets of North America and Europe. This phenomenon, together with the substantial fall in the value of the rouble and the hryvnia during the summer, were two of the essential factors in the commercial problems reported in the steel sector in 1998.

Crude steel production

- World: World crude steel production fell by 2.8% in 1998 to 776.2 million tonnes, 14 million tonnes less than the record level achieved in 1997.

- OECD: Crude steel production for the area as a whole fell by 3.3% in 1998, 16 million tonnes down on 1997. Output totalled 466.1 million tonnes, rising in Australia and Canada, but falling in other Member countries. In Brazil, too, crude steel production fell by 1.5%.

- Crude steel production fell in nearly all market economy countries outside the OECD area, with the exception of Chinese Taipei. In the ASEAN(5) countries, production dropped by 17.5%.

- In the NIS overall, crude steel production fell by 8.1%. In Russia, production declined by a further 9.6%, while in The Ukraine it fell by only 4.6%.

- In the countries of Central and Eastern Europe, steel production fell by 8.3%.

- In China, crude steel production reached another record level of 114.4 million tonnes, an increase of 5% on 1997. For the third year running, China was the world's leading steel producer.

Steel capacity utilisation rate

- In the OECD area overall, the average capacity utilisation rate declined to an average of 78% in 1998.

- Capacity utilisation rates were 86% in the United States, 81% in the European Union and only 62% in Japan.

- In most other areas of the world, with the exception of the NIS and the countries of Central and Eastern Europe, steel production capacity continued to increase but, as a result of the crisis in Asia, at a significantly slower pace. In China, capacity utilisation exceeded 92%, while worldwide the average capacity utilisation rate in 1998 was just over 74%.

Steel prices

- The disruption in the world steel market prompted a substantial decline in steel product prices over the year. This fall in prices was perceptible in all markets. The decline in prices, down by 40%, or even more in the case of certain flat products, appears to have halted around November 1998.

1999

Apparent steel consumption

- World: Apparent steel consumption worldwide could fall by just under 3% in 1999 compared with 1998, *i.e.* by some 20 million tonnes.

- OECD: The OECD area is not expected to escape the worldwide decline and steel consumption in the area overall could fall by just over 3.8%, *i.e.* by 15.6 million tonnes on 1998 levels. Trends will be downward in all Member countries, apart from Mexico and Korea.

- In the United States, steel consumption should decline by 9%, it should fall by 5.3% in the European Union, by 3.5% in Canada and by about 1% again in Japan. In Brazil, too, a substantial decline is expected of around 8.5% in the demand for steel.

- The demand for steel should continue to fall in all market economy areas outside the OECD, but could begin to pick up slightly in India and certain other Asian countries. In the ASEAN(5) countries, however, the fall in consumption should slow down to a mere 4.5%.

- In the countries of Central and Eastern Europe, the decline in steel consumption should continue and could fall by nearly 7%.

- In the NIS, the demand for steel seems likely to recover in 1999.

- In China, the demand for steel is likely to remain high in 1999, exceeding 100 million tonnes, but it could decline by around 2% on 1998 levels.

Steel trade

- In volume terms, world steel trade is expected to be 11.8% down on 1998, a fall of almost 21 million tonnes. This decline will be due largely to the normalisation of trade flows as a result of the fall in steel consumption in the American and European markets.

- Total net exports in the OECD area should begin to pick up, reflecting a decline in steel trade, with steel exports falling slightly less than imports.

- Net exports from the European Union are expected to rise, with decline in domestic demand accompanied by a 20% fall in imports, while exports should stay close to their 1998 level. Conversely, in Japan, net steel exports are expected to decline but this time owing to a decline in both exports and imports, caused by a slowdown in the domestic market.

- In the United States, net steel imports should fall by about 19%. The decline in steel exports is expected to continue, but imports may well fall by 7 million tonnes. Net exports from Korea could decline by nearly 35%, with the upturn in consumption causing a reduction in exports together with a slight increase in imports.

- Net imports from market economies in general outside the OECD area should continue to fall by around 7.5%. This trend should apply to all areas other than Latin America, where net steel imports should begin to rise again.

- Net steel exports from the NIS are expected to fall by around 2.5 million tonnes, 8% down on 1998.

- Chinese net imports should continue to rise as a result of the fall in exports, particularly those to other Asian countries.

Crude steel production

- World: As a result of the slowdown in world demand, crude steel production should also decline by around 2.9% in 1999, *i.e.* by some 22 million tonnes compared with 1998 levels.

- OECD: Crude steel production in the area as a whole should fall by 3.5%, with production declining in nearly all Member countries other than Canada and Mexico. In Brazil, too, crude steel production could fall by around 4.5%.

- While the downward trend is expected to continue in the ASEAN(5) countries, Latin America and Africa, production could begin to recover slightly in the Middle East and the upward trend should continue in Chinese Taipei.

- In the countries of Central and Eastern Europe that are not OECD Members, crude steel production should continue to decline significantly, *i.e.* by around 10%.

- As for the NIS, the recovery in domestic demand should result in higher production, which may rise by just over 2%.

- In China, the government wishes to reduce crude steel production in 1999 and implement a vast restructuring and reorganisation plan for its steel industry. It seems unlikely, however that its goal of a 10% reduction in output can be achieved. At best, production could fall by around 3.5%.

Crude steel making capacity utilisation

- In 1999, crude steel making capacity should continue to increase by around 3% worldwide. However, it is already clear that a number of projects planned for the future have been either cancelled or postponed as a result of the Asian crisis.

- In the OECD area, production capacity in 1999 is unlikely to be up on 1998, but the fall in steel production should bring the average capacity utilisation rate down to 76%.

- In China, production capacity should rise by a further 5.2%, or 6.5 million tonnes per year, but the capacity utilisation rate should fall from 92% in 1998 to 84% in 1999. In the other regions, particularly in Asia, the capacity utilisation rate should fall substantially, possibly to around 33% in certain countries.

Steel prices

- Throughout the first quarter of 1999, steel prices remained at the very low level they had fallen to at the end of 1998. However, as the stocks accumulated in 1998 are reduced over the first half of the year and demand in several markets gradually picks up over the second half, prices may well increase as from the third quarter.

2000

Apparent steel consumption

- World: Following the expected decline in world demand for steel in 1999, apparent steel consumption may recover slightly in 2000, rising by 4% or even more.

- OECD: With economic growth in 2000 estimated at 2.3% for the area as a whole (the growth should be driven by private consumption and investment, and this could apply in all Member countries), apparent steel consumption in the OECD area may well begin to recover, increasing by just over 3% in 2000.

- In Brazil, the demand for steel should begin to increase substantially in 2000, rising by more than 4.5%.

- In the countries of Asia, including the ASEAN(5) group, steel consumption should begin to rise after the sharp declines in certain countries in 1997, 1998 and 1999.

- In the NIS, the recovery in steel consumption, which should become apparent in 1999, is expected to gather pace in 2000, particularly in The Ukraine. However, it should be noted that the increase of around 10% in steel consumption, forecast for the NIS overall, will only bring consumption to 22% of the record volume set by the USSR in 1987.

- In China, the demand for steel could restart to pick up in 2000 and stabilise at much the same level as in 1998, levelling off before increasing in subsequent years.

Steel trade

- World trade in steel could remain at much the same level as in 1999, accounting for about 23% of apparent steel consumption worldwide. This is the likely outcome of commissioning additional capacity in different parts of the world and of the recovery of consumption in these different regions.

- Total net exports from the OECD area should again be higher than in 1999, but the increase would be due chiefly to a decline in imports with only a marginal increase in exports. While net exports from Japan are expected to fall slightly, those from the EU(15) should increase in spite of the recovery in domestic demand. In contrast, a further substantial decline in net imports is likely in the United States, where the market share of imports should be around 26%.

- Steel imports should begin to pick up slightly in Latin America and the Middle East and should rise in the Asian countries.

- Net exports of steel from the NIS are expected to continue to decline, but those from the countries of Central and Eastern Europe could increase, following the upturn in demand in the other European countries.

- In China, steel imports should decline, while exports are expected to pick up again.

Crude steel production

- World: As a result of the expected recovery in steel consumption, crude steel production should also increase by just over 4% in 2000.

- OECD: Steel production in the area as a whole could increase by about 3.8%, or 17 million tonnes more than in 1999.

- Production is expected to increase among Member countries generally, apart from Australia and Canada.

- In Brazil, crude steel production could increase by about 2.9%, exceeding 25 million tonnes once again.

- Crude steel production is expected to pick up again in the NIS and China, and it should remain on an upward trend in all other regions.

Steel production capacity utilisation

- Crude steel production capacity in the OECD area as a whole is expected to increase by a further 4 million tonnes in 2000, accounting for 54% of world capacity. The average capacity utilisation rate should be in the region of 78%.

- In all the other areas, steel production capacities will continue to increase but at a slower rate, which should lead to widespread improvements in utilisation rates provided that the expected recovery in output materialises.

Steel prices

- As forecast, steel market trends in 2000 should cause the price of steel products to rise at a faster rate.

Graph 2. **OECD apparent steel consumption, 1970-2000**
Million tonnes of product equivalent

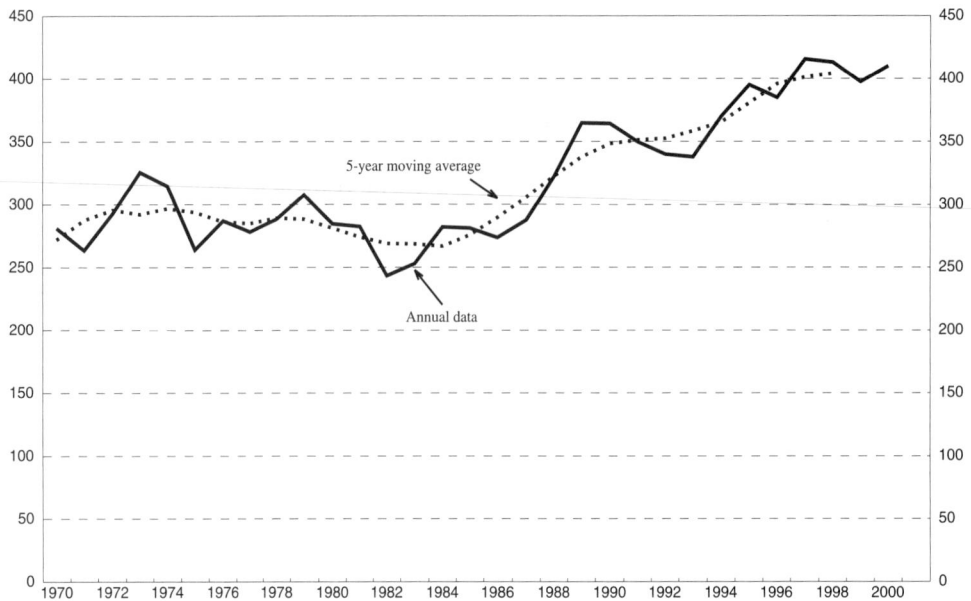

5-year moving average

Annual data

Source: OECD.

Graph 3. **OECD crude steel capacity and production, 1970-2001**
Index, 1973 = 100

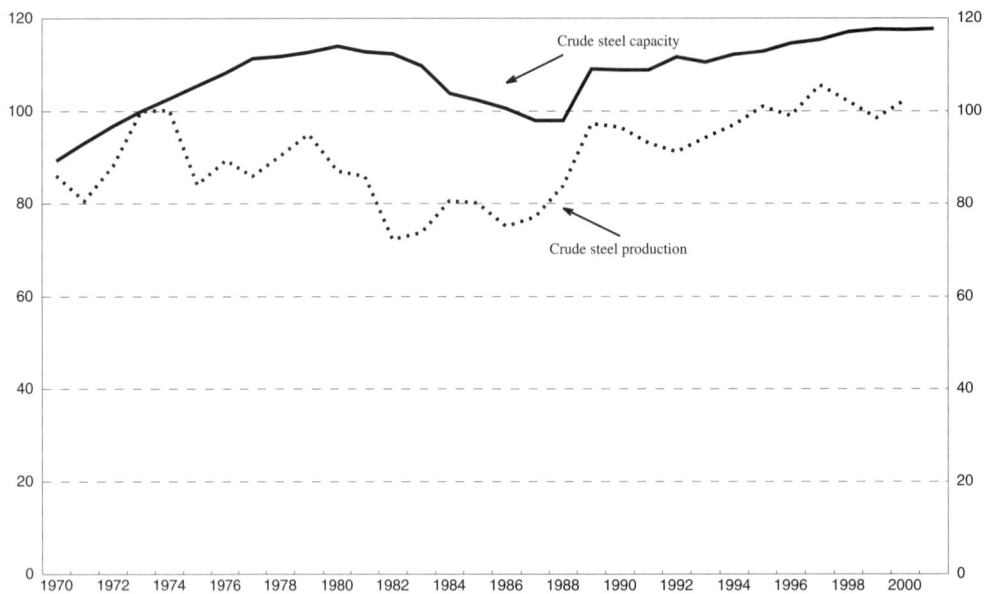

Crude steel capacity

Crude steel production

Source: OECD.

18

Main results

	1998 variation in		1999 variation in		2000 variation in	
	million tonnes	%	million tonnes	%	million tonnes	%

a) **Change in apparent steel consumption (in product equivalent)**

United States	+9.3	+8.6	-10.5	-8.9	+3.2	+3.0
Japan	-14.0	-17.2	-0.7	-1.0	+1.9	+2.8
EU(15)	+11.2	+8.5	-7.3	-5.1	+2.2	+1.6
Other Europe	+1.9	+7.1	-0.2	-0.5	+0.9	+3.3
Canada	+1.5	+9.3	-0.6	-3.5	+0.7	+4.1
Korea	-14.0	-37.1	+3.4	+14.3	+2.9	+10.8
Mexico	+1.7	+22.0	+0.3	+2.8	+0.7	+7.0
Oceania	0.0	0.0	-0.0	-0.2	+0.2	+2.6
Total OECD	-2.5	-0.6	-15.5	-3.8	+12.6	+3.2
Brazil	+0.5	+3.3	-1.3	-8.5	+0.6	+4.7
Central and Eastern Europe	-0.2	-4.4	-0.3	-6.9	+0.4	+9.7
NIS	-2.1	-8.7	+3.9	+17.6	+2.5	+9.5
China	+8.2	+8.4	-2.1	-2.0	+2.1	+2.0
Rest market economies	-19.4	-14.1	-4.3	-3.6	+8.7	+7.6
World	-15.8	-2.3	-19.7	-2.9	+27.3	+4.1

b) **Net trade (in product equivalent)**
See Section *d)* below for detailed import and export figures

EU(15) net exports	-11.2	-95.5	+4.7	+890.6	+1.8	+33.3
Japan net exports	+3.8	+23.2	-4.8	-23.6	-0.2	-1.1
US net imports	+10.1	+43.3	-6.3	-18.8	-2.2	-8.1
Rest OCDE net exports	+5.7	+51.8	-5.2	-31.1	-1.6	-13.9
Total OCDE net exports	-11.8	-74.7	+1.1	+27.5	+2.2	+43.1
China net imports	+2.5	+51.0	+1.1	+14.9	-2.6	-30.6
Other economies net imports	-13.2	-68.0	-0.1	-1.6	+6.9	+113.1

c) **Crude steel production**

United States	-0.8	-0.9	-4.9	-5.0	+6.3	+6.8
Japan	-11.0	-10.5	-5.8	-6.2	+1.8	+2.1
EU(15)	-0.2	-0.1	-2.8	-1.8	+4.4	+2.8
Other Europe	-1.8	-4.8	-1.7	-4.9	+2.6	+7.7
Canada	+0.4	+2.5	+0.3	+1.7	0.0	0.0
Korea	-2.7	-6.2	-1.4	-3.4	+1.6	+4.2
Mexico	-0.0	-0.3	+0.9	+6.3	+0.9	+6.0
Oceania	+0.1	+1.1	-0.7	-7.2	-0.7	-7.8
Total OECD	-16.1	-3.3	-16.1	-3.5	+17.0	+3.8
Brazil	-0.4	-1.5	-1.2	-4.5	+0.7	+2.9
Central and Eastern Europe	-1.1	-8.3	-1.2	-10.1	+1.6	+14.9
NIS	-6.6	-8.1	+1.6	+2.2	+2.2	+2.9
China	+5.4	+5.0	-4.2	-3.6	+5.5	+5.0
Other market economies	-3.6	-4.1	-1.2	-1.4	+4.8	+5.8
World	-22.7	-2.8	-22.2	-2.9	+31.7	+4.2

Main results *(cont.)*

	1998			1999			2000		
	Imports	Exports	Net trade	Imports	Exports	Net trade	Imports	Exports	Net trade
d) Imports, exports and net trade (in million tonnes)									
EU(15)	23.4	24.0	-0.5	18.8	24.0	-5.3	18.5	25.5	-7.0
Japan	4.8	24.9	-20.1	4.2	19.6	-15.4	4.0	19.2	-15.2
United States	38.3	5.0	33.3	31.4	4.4	27.0	29.4	4.6	24.8
Rest OECD	30.5	47.3	-16.8	28.5	40.0	-11.5	29.3	39.1	-9.8
Total OECD	97.1	101.7	-4.0	82.8	87.9	-5.1	81.1	88.4	-7.3
NIS	4.6	38.0	-33.4	4.7	35.5	-30.8	4.5	34.5	-30.0
Central and Eastern Europe	1.6	6.5	-4.9	1.4	5.6	-4.3	1.4	6.5	-5.1
China	13.1	5.7	7.4	13.0	4.5	8.5	11.4	5.5	5.9
Other market economies	64.1	27.4	36.7	57.3	24.3	33.0	61.5	23.6	37.9
World	180.4	178.7	1.7	159.2	157.8	1.5	159.9	158.5	1.4

	Capacity in million tonnes			Utilisation rate in %		
	1998	1999	2000	1998	1999	2000
e) Crude steel production capacity[1] and utilisation rates						
United States	113.3	116.3	116.3	85.9	79.8	85.2
Canada	16.7	16.7	16.7	94.8	96.4	96.4
Korea	42.5	42.5	42.5	93.9	90.7	94.5
Mexico	16.5	17.0	18.0	86.1	88.8	88.9
Oceania	10.0	9.6	8.3	97.0	93.8	99.4
Japan	149.6	150.0	150.0	62.6	58.5	59.7
EU(15)	198.2	197.8	198.2	80.7	79.3	81.3
Other Europe	44.4	44.7	44.7	80.2	75.8	81.6
Total OECD	591.2	594.6	594.7	78.8	75.7	78.5
Brazil	31.2	29.9	31.1	82.7	82.3	81.4
NIS	141.3	141.9	141.9	52.7	53.6	55.1
Central and Eastern Europe	20.2	20.2	21.1	59.4	53.4	58.8
China	124.2	130.7	131.8	92.1	84.3	87.8
Other economies	132.8	153.9	176.5	62.8	53.4	49.3
World	1 040.9	1 071.2	1 097.1	74.6	70.4	71.6

1. Estimations of capacity for OECD countries are based on the Steel Committee's Annual Survey of Effective Capacity. There are differences in country definitions, and the changes in a country's operating rates from year to year are more significant than direct comparisons between the various countries' operating rates.

DEVELOPMENTS IN THE STEEL MARKET BY AREA

United States

Surprisingly, economic growth in the United States in 1998 remained very strong. GDP rose by 3.9% over the year, the strongest increase in the past ten years, and by as much as 6% at annual rates in the fourth quarter, the highest level in 15 years. Private consumer spending, which accounts for some two-thirds of GDP, increased by 4.9% over the year, reflecting a high level of consumer confidence in the US economy, an increase in disposable income and the positive impact of low inflation. Investment in non-residential construction, up by 10.3% on 1997, also continued to boost the economy. Low interest rates helped the construction sector and investment in the residential construction sector jumped by 10.4% over the year. Throughout 1998, 2.75 million jobs were created and the average unemployment rate was 4.5%, the lowest since 1970. Despite strong growth and low unemployment rates, inflation remained under control. Wholesale prices were down by 0.1%, mainly as a result of a 12.1% drop in energy prices.

Although economic activity overall remained very buoyant, the manufacturing sector suffered from losses in its share of the export market and from a substantial increase in cheap imports. Output in the automobile sector was 11.6 million units in 1998, down by 1.1% on 1997. The construction sector became the principal customer of the US steel industry in 1998. Other growth areas in 1998 were the agricultural and capital goods sectors; however, activity declined in such sectors as oil, gas and petrochemicals, containers and metal processing.

Production indices, 1995 = 100 (seasonally adjusted)

	1996	1997	1998	1998 Fourth quarter	1998/97 % change
Industrial production	104.5	110.8	114.8	115.1	3.6
Manufacturing industries	104.8	111.9	116.6	117.4	4.2
Motor vehicles and parts	99.4	105.4	106.4	108.8	0.9
Fabricated metal products	103.3	107.1	109.3	110.4	2.1
Non-electrical machinery	103.7	108.9	112.8	110.9	3.5
Electrical machinery	125.4	157.2	182.8	197.4	16.3
Mining	101.7	103.8	102.1	99.5	-1.6

Source: OECD, *Indicators of Industrial Activity*.

In 1999, economic growth may well slow down slightly and GDP could grow at a rate of around 3%. This slowdown will reflect the adverse impact of the build-up of stocks in 1998. Overall investment could also decline, although a reduction in the price of a wide range of capital goods has made capital expenditure more attractive. As regards residential investment, the upward trend recorded during the fourth quarter of 1998 seems bound to continue into 1999 and is unlikely to slow down until the second half.

The rise in private consumer spending in 1999 is not likely to continue at the same rate as in 1998. Further vigorous growth is likely in the automobile market, but orders for durable goods are expected to slow down, since demand is declining in the aeronautics, electronics and capital goods sectors. The unemployment rate may well continue to fall and inflation should remain under control at between 2% and 2.5%.

In 2000, the forecasts currently available suggest that the growth of the US economy could revert to its long-term trend, showing an increase of between 2% and 2.5%. Private consumer spending should again increase by 2.5%, with investment increasing by a further 2.2%. The growth in industrial output should begin to pick up slightly, levelling out at 2.6%.

Benefiting from the continued economic growth in 1998, the steel market again registered record levels of demand. Apparent steel consumption, expressed as finished product equivalent, reached a record 117 million tonnes, up 8.6% on 1997. However, the market experienced an unprecedented increase in cheap imports, and US steelmakers' deliveries to the domestic market consequently fell by 3.5%, as the share of imports in the US market increased substantially. Deliveries to the construction sector increased by 1%, making it the principal customer of US steel makers, but deliveries to the automobile sector declined by 2%, and deliveries to the processing and key oil sectors fell by 20.1%, and 33.6% respectively.

The price of steel products fell steeply, in some cases to their lowest level in years. Between May and December 1998, the price of hot-rolled, cold-rolled and galvanised sheet metal fell by an average of 21%. The compound price index for steel products as a whole fell by more than 7.5%. With price reductions of this order, together with lower delivery volumes, the profitability of firms suffered severely. As a result, employment also fell steeply over the second half of the year.

Crude steel production increased by 6.7% over the first half of 1998 compared with the first half of 1997, but it fell sharply in the last months of the year and remained at 97.7 million tonnes over the twelve months of 1998, down by 0.9% on 1997. However, a 4 million tonne increase in crude steel production capacity caused the average capacity utilisation rate to fall to 85.9% in 1998, compared with 90.1% in 1997. Here too, the capacity utilisation rate had been nearly 92% from January to June 1998 before falling below the 75% mark during the last two months of the year. In 1998, continuously cast steel production continued to account for nearly 94.5% of all US steel production.

As regards trade in steel products, with high domestic demand and the rise of the dollar against the currencies of other steel-producing countries, steel imports rose by 33.3% to a record 37.7 million tonnes, 9.4 million tonnes more than in 1997. The rate of import penetration on the US market went up from 26.7% in 1997 to 32.7% in 1998, another record level. Imports from three countries, Japan, Russia and Korea, accounted for 75% of this increase. Steel imports from Japan jumped by 163% to 6.1 million tonnes and imports from Korea increased by 109% to 3.1 million tonnes. As a result of the crisis in Asia and the fall in demand in that area, Russia diverted some of its exports to the United States, increasing them by 59% to 4.8 million tonnes. Imports from a certain number of other countries: Australia, South Africa, The Ukraine, China, Chinese Taipei and Indonesia also rose substantially. Imports from the EU, however, were down by 3.6%, falling to 6.4 million tonnes. The fall in the price of imports contributed to a further destabilisation of the US market.

As a result of the fall in world prices and the low level of overseas demand, steel exports from the United States fell by 8.5% to 5 million tonnes in 1998. Exports to Canada were down by 10%, amounting to a mere 3 million tonnes, whereas exports to Mexico increased by 8% to 1.1 million tonnes. These two countries alone absorbed more than 81% of all US steel exports. Exports to the EU were down by 16%, amounting to only 0.2 million tonnes.

Given that the US economy is expected to be strong in 1999, steel demand should remain high but apparent consumption could slow down as a result of the slow rate of stock decumulation. Apparent steel consumption could fall by nearly 8.5%, *i.e.* 10 million tonnes down on 1998, but real consumption would only decline by around 5%. Imports should fall quite significantly, by around 16.5%, as a result of low prices in the US market, fall in demand, reduction in stocks and the impact of the antidumping measures applying to a certain number of countries. The fall in world demand should lead to a subsequent reduction in steel exports, which could fall by 12%. Production capacities will increase by a further 3 million tonnes. But the low level of demand and the reduction in stocks should entail a 5% fall in crude steel production which would thus amount to 92.8 million tonnes. The rate of capacity utilisation is expected to decline to around 80%. The profits of US producers should be at their lowest during the first quarter of 1999, but should gradually improve once prices have begun to pick up, probably at the end of the second quarter of 1999.

As for 2000, the steel market in the United States should begin to recover, with apparent consumption increasing by around 3%. Crude steel production should increase more substantially, by around 6.5%, to 99 million tonnes. Net steel imports by the United States are expected to fall by around 8%, reflecting a decline in imports of 6% or more and a slight recovery in exports due to an improvement in world steel demand. The rate of import penetration on the US market could fall to 26.5%, roughly equivalent to the 1997 level, whereas the capacity utilisation rate should increase again to over 85%.

Canada

Economic growth in Canada slowed down slightly in 1998 and the GDP growth rate stood at 2.8%, compared with 4% in 1997. Inflation remained low at 0.9% even though the unemployment rate fell to an average of only 8.3% in 1998 compared with 9.2% in 1997. Interest rates rose slightly to 6.75% by the end of 1998 compared with 6% the previous year. Canada's economic growth continues to be driven by exports and in this respect the relatively low value of the Canadian dollar has served both to increase exports and to make imports less competitive.

Industrial output was up by 2.4%. In the automobile sector, vehicle production fell by 2.3% to 2.35 million units in 1998, but at the same time, new vehicle sales rose by 0.3% to 1.43 million units, the highest level since 1989. In the construction sector as a whole, activity was up by 7.2%, rising by 20.1% in the non-residential sector but falling by 2% in the residential sector. In the pipe sector, activity fell by 0.6%, and steel deliveries fell by 16.8% in 1998.

Crude steel production in Canada increased by 2.5% in 1998 to 15.8 million tonnes, the highest level since 1980, 98.5% of which was continuously cast steel. Total steel imports continued to rise in 1998, owing to ever-increasing demand, up by more than 17% on 1997 to 7.4 million tonnes, an all-time record. Likewise, the rate of import penetration in the Canadian market rose by 13.4% to account for 41.9% of the market. Imports from Asia increased substantially, particularly those from Japan, Korea, India, Chinese Taipei and Indonesia, as did imports from Russia, Romania, Slovakia and, to a lesser extent, from the EU, Australia, South Africa and Turkey. On the other hand, there was a decline in steel imports from the United States and Central and South America. The greater part of these imports were concentrated on a few product categories, mainly hot-rolled and cold-rolled sheet metal, hot-rolled coils and H-sections.

In the same period, steel exports increased by 0.6%, barely more than 4.2 million tonnes. In 1998 total steel demand in Canada was again up on the record level of 1997 and apparent steel consumption, expressed in finished product equivalent, rose by 9.3%. Despite the strength of the market, the income of Canada's leading steel manufacturers fell in 1998.

Projections for the Canadian economy in 1999 and 2000 forecast steady growth, though at a slightly lower rate, and GDP could be up by around 2.9% in 1999 and 2.8% in 2000. Inflation should remain under control, although it could creep back up to 1.1% in 1999 and 1.7% in 2000. The economic trends should also allow a reduction in the unemployment rate to 8.1% in 1999 and 7.9% in 2000. Investment should continue to rise steadily, by between 6% and 7%, and private consumption should also continue to rise but at a slower rate, given the expected increase in interest rates – up by an average of 6.6% in 1998, 7.6% in 1999 and 8% in 2000 – and in the value of the Canadian dollar against the US dollar.

The increase in industrial output is likely to continue, with further rises of 3.2% in 1999 and 3.7% in 2000. Production in the automobile sector should remain strong, given that demand for replacement vehicles is expected to be high until the year 2000. In the residential construction sector, activity is expected to remain flat in 1999 and to pick up by around 5.8% in 2000.

The steel market will remain very buoyant in 1999, though apparent steel consumption should fall by 3.5%, mainly as a result of the stock adjustment process in the first half of 1999. Apparent consumption should start to pick up again over the second half of the year. In 2000, apparent steel consumption is expected to rise again, growing by nearly 4%. In 1999, steel imports should fall quite significantly – perhaps by more than 11% – as a result of the slowdown in consumption, but they should start to pick up again in 2000 if the rate of consumption increases as foreseen. Steel exports are expected to decline, just slightly in 1999 and then more substantially in 2000. Crude steel production may well reach a record level of 16.1 million tonnes in 2000.

Mexico

In Mexico, following the record economic growth level reported in 1997, real GDP was up again by 5% in the first half of 1998, sustained by investment and private consumption. But the more restrictive economic measures introduced to counter the effects of the decline in the price of oil and the fall in the value of the peso slowed down economic activity in the second half of the year. In the course of 1998, GDP increased by 4.6%. Investment, however, rose by 10% and private consumption by 6%. Overall domestic demand increased by a further 5.7%, but imports slowed down and exports of manufactured goods increased over the first half, limiting the deterioration in the balance of trade. The improvement in the work market continued and salary increases of around 18% were negotiated, with the result that buying power increased for the first time since the crisis of 1994/95. Nominal interest rates were substantially raised at the end of summer and should remain high in 1999, prompting a downturn in GDP, which is not likely to rise by more than about 3.6% with private consumption falling off. Forecasts for 2000 point to a probable recovery in investment in industry, a fall in inflation, and a rise in private consumption, with a consequent increase in GDP of around 4.4%.

After the strong recovery in apparent steel consumption reported in 1996 and 1997, demand for steel continued to rise by nearly 22% in 1998, or 1.7 million tonnes more than in 1997. To meet this demand, steel imports rose by around 40% and at the same time exports fell by 17.5%. Crude steel production remained close to the 1997 level at 14.2 million tonnes, and 86% of output was continuously cast. With production capacity up to 18.1 million tonnes, the capacity utilisation rate was 79%.

Steel demand is expected to remain steady in 1999, with an increase of no more than 2.8% on the record 1998 level. Steel exports should continue to decline by around 16.2%, while imports should fall by more than 55% owing to the weakening of the national currency. The production of crude steel should continue to increase, rising by 6.3% to 15.1 million tonnes. In 2000, the increase in steel demand may well continue, and substantial growth of around 7% is expected, which means that steel imports will probably increase too, by around 20%, particularly if exports also start to rise as a result of the expected recovery in world demand. Crude steel production could therefore rise by 6% to 16 million tonnes. As new production capacity starts to come on stream, the average utilisation rate of production capacity may well rise to 85% in 2000.

European Union (15)

In 1998, the EU economy held out well against the worsening international situation and growth in GDP for the EU as a whole was 2.9%. This growth was supported by strong domestic demand, which compensated for the fall in foreign demand, resulting from the poor international climate. Private consumption rose by 2.5%, and the rate of job creation by 1.2%. The outlook for unemployment improved and the average level in the EU fell to 10%. Investment, which had started to pick up in 1997, gathered pace in 1998 and rose by 5%. The average rate of inflation was down further, amounting to just 1.8% in 1998.

The buoyancy of economic activity helped to fuel strong growth across the manufacturing sector, in response to an increase in domestic demand, particularly during the first half of the year. The repercussions of the crisis in Asia had spread to other parts of the world, and became more marked in the EU towards the middle of 1998. The vigorous activity noted in the steel-consuming sectors in the first half of 1998 subsided towards the end of the year, except in the construction sector, which showed a slight improvement throughout the whole of 1998. Industrial output rose by an average of 4% in the EU(15) as a whole. The automobile sector benefited most substantially from this growth, with activity up by 9%.

Production indices, 1995 = 100 (seasonally adjusted)

	1996	1997	1998	1998 Fourth quarter	1998/97 % change
Industrial production					
Germany	100.6	104.1	108.5	111.7	4.2
Spain	99.3	106.1	111.8	109.1	5.4
France	100.2	104.1	108.8	109.0	4.5
Italy	97.1	99.8	100.6	102.7	0.8
United Kingdom	101.0	101.8	102.7	106.0	0.9
EU(15)	100.1	104.5	108.0	107.6	3.3
Metal-using industries					
EU(15)	98.9	102.9	107.3	108.6	4.3
Of which:					
Motor vehicles	103.5	110.5	122.9	129.4	11.2
Mechanical engineering	100.9	104.2	107.5	113.7	3.2

Source: OECD, *Indicators of Industrial Activity*.

The difficult global context, which caused economic growth in the EU to slow down during the second half of 1998, should result in a further reduction in the growth rate in 1999, due to the effects of international problems and their impact on trade volumes and investment. Forecasts have been revised downwards and GDP growth in 1999 is expected to be no more than 2.1%, rather than the previously forecast 2.4%. These factors, combined with political uncertainties, should have a more adverse effect on the growth of some of the major EU economies, particularly those of Germany and Italy. A gradual improvement in the economy can nevertheless be expected during the second half of 1999. Leaving aside these largely external factors, there can be no doubt that the fundamentals of the EU economy are sound: inflationary pressures are weak, interest rates very low and consumer confidence suggests that the slowdown in growth should very soon be over. Overall domestic demand should increase by around 2.5% in 1999 and investment should be up by only 3.7%, and yet inflation could be particularly low, falling to an average of 1.3%, and unemployment should continue to fall, probably to around 9.6%.

Growth in output in the main steel-consuming sectors is expected to fall off in 1999, especially if compared with the very high growth rates reported in 1998. This would be due to reduced industrial activity over the first half of the year. If the international situation were to improve over the second half of 1999, investment could pick up. The situation should vary somewhat from one sector to another, and output should remain buoyant in those with a particular orientation towards the domestic market. A clear improvement is accordingly expected in the construction sector, both residential and non-residential, after a number of flat years. In the automobile sector, however, after several years of vigorous activity, output can at best be expected to remain stable compared with 1998. In the mechanical engineering and preliminary metal processing sectors, the slowdown in industrial output should make a more marked impression. These sectors are highly responsive, however, and a recovery in investment before the end of 1999 could result in an increase in output before the end of the year.

Economic growth should be more vigorous in 2000 than in 1999 and GDP should rise by an average of 2.7% in the EU. Growth in investment should accelerate and might well increase by 4.8%. Private consumption should also continue its upward trend with growth expected to be in the region of 2.8%. The average rate of unemployment in the EU(15) area should continue to decline, so that no more than 9.2% of the working population are hit. The upward trend in the level of activity in the main steel-consuming sectors should remain unchanged, as in 1999.

In the steel sector, after a steep increase in 1997, apparent steel consumption in the EU(15) area (in finished product equivalent) rose again by 8.5% in 1998, an increase of some 11.2 million tonnes. The real consumption of steel rose less, by only 2%, as a result of a substantial increase in stocks held by producers, merchants and consumers.

In 1998, crude steel production in the EU as a whole was slightly down on the record 1997 level, amounting to only 159.6 million tonnes. Production declined in Germany, Italy, the United Kingdom, Luxembourg, the Netherlands and Portugal, the rate of decline varying from one country to another, but rose in all the other countries. In 1998, EU steel trade was seriously affected by disruption on the international scene. Imports increased by 43% while at the same time exports declined by 15%. The combined effects of the substantial rise in imports and the significant fall in exports severely upset the net balance of trade in steel. The balance, which had always shown a surplus in the past, consequently fell by 11.7 million tonnes in 1997 to just 0.5 million tonnes in 1998, *i.e.* the loss amounted to 11.2 million tonnes.

Yearly percentage changes in real and apparent steel consumption in the EU area

	1998/97 Realised		1999/98 Estimates		2000/99 Forecast	
	Real	Apparent	Real	Apparent	Real	Apparent
Germany	3.7	3.9	-7.1	-8.5	-0.5	-0.8
Spain	5.5	12.0	13.4	2.7	2.6	0.0
France	11.2	17.6	1.6	-3.6	-4.7	-3.8
Italy	-1.1	4.8	2.7	-3.0	4.9	5.9
United Kingdom	0.7	3.6	3.7	-3.5	5.7	4.3
Rest EU(15)	5.2	13.7	4.6	-7.4	3.3	3.1
Total EU(15)	2.0	8.5	3.1	-5.1	1.6	1.6

Source: OECD.

In 1999, demand for steel within the EU area is expected to slow down, mainly at the beginning of the year, but it could subsequently pick up and apparent consumption could fall by around 5.1% over the year, largely because of the need to decumulate stocks, which will become apparent mainly during the first six months. Real steel consumption, on the other hand, should remain at a high level. Crude steel production could fall by 1.8% to some 156 million tonnes. The utilisation rate of production capacity should decline slightly to about 79%. In this context of falling demand and significant reduction in stocks, EU imports should show a decline in the region of 20%. Steel exports from the EU, however, will probably stabilise to some extent as there are no clear signs of improvement on the international markets. As a result of these developments, the balance of trade in steel products could improve, rising again to some 5.3 million tonnes. As for developments in the prices of steel products, an improvement is likely, which should become most pronounced during the second half of the year, after a period of stagnation at floor prices in the last months of 1998 and the first months of 1999.

Apparent steel consumption should begin to pick up by around 1.6% in 2000, by slightly over 2 million tonnes on 1999. This increase in demand in the domestic market, accompanied by a slight drop in imports (-1.3%) and a gradual improvement in exports (+6.3%), could prompt an increase in the production of crude steel of perhaps 2.8%. As a result, the production of crude steel in the EU would then exceed the 161 million tonne mark, and the capacity utilisation rate would be higher than 81%.

Other European countries

This area comprises the Czech Republic, Hungary, Norway, Poland, Switzerland, Turkey and the former Yugoslavia. After the sharp increase in consumption reported in 1997 (+14.2%), demand in this area continued to rise, up by 7.1% in 1998, *i.e.* 2 million tonnes more than in 1997. Despite a significant increase in the market, the sharp fall in net steel exports led to a 4.8% decline in the production of crude steel in these countries as a whole. Steel imports increased by 6.8% and at the same time exports fell by 11.5%, as a result of which net exports from the area were down by more than 70%.

In 1999, steel demand is expected to decline slightly, owing to the impact of more serious problems at international and European level. Apparent steel consumption could fall by around 0.5%. Net exports should again decline substantially, by around 87%, to practically zero, as imports stabilise at a level close to that achieved in 1998 and exports fall by more than 10%. Crude steel production should decline by nearly 5%.

The year 2000 should see steel consumption picking up throughout this area, with further growth of around 3.3%. This increase will probably be accompanied by a rise in crude steel production of around 7.7% and an increase in net exports from these countries, which would mean a 2.8% decline in imports and more especially a recovery in exports of around 6%.

In **Hungary**, the upward trend of the economy continued in 1998, with an increase in GDP of around 5.5% and improvement in all the main economic indicators. Inflation slowed down significantly, apparently not exceeding 14.3% over the year, and salaries increased by an average of 18.8% so that buying power rose by 4%. The economy was stimulated by a boost in industrial output, which has been growing strongly in recent years, and was up by a further 13% in 1998. The machinery and plant sector made the major contribution to this growth. Despite these very positive developments, certain external factors tended to slow the rate of growth over the second half of the year: the economic and financial crisis in Russia, increasing uncertainty about the "emerging" markets and the slowdown in growth on the most important of Hungary's western markets. In 1999, growth in GDP should be between 3% et 4%, inflation should slow down to just 9% and buying power could increase by a further 5%. Industrial output should increase by 6% to 7% and investment should continue to rise by around 9%, resulting in a strong increase in activity in the construction and mechanical engineering sectors. On the whole, despite a slight reduction on 1998 levels, the strong upward trend in economic activity should continue in 1999. This trend may well be maintained in 2000, with GDP growth rising to 3.2%. Inflation should continue to fall and investment should continue to rise by around 6%. Industrial output should increase by 7%.

Apparent steel consumption increased very rapidly in 1998, up by 21.3% on 1997. Despite growing domestic demand and a slight increase in steel exports, crude steel production only increased by 7.7%, the principal beneficiaries of increased demand being foreign suppliers, who provided for around 70% of the increase in consumption. The firm DUNAFERR, which produced 80% of Hungary's total steel output, mainly produces hot- and cold-rolled flat products, and was unable to satisfy the demand for long products in the construction sector, which experienced strong growth in 1998. The gap between the increase in steel consumption and steel output thus prompted a strong rise in imports of steel – around 22.8% – to cope with demand. It should be noted that steel imports rose by nearly 95% between 1996 and 1998, which led the Hungarian Government to reintroduce measures to protect the domestic market from the influx of certain product from Russia and The Ukraine over the second half of 1998; in addition, negotiations were held with other countries with a view to keeping imports at a reasonable level.

In 1999, although a upward trend is still expected, it is likely to be less marked than in the past. A further increase of around 5.3% is expected in steel demand, accompanied by an increase of just over 1.5%, in crude steel output, all of which should be continuously cast. Steel imports should continue to rise at the lower rate of 2.7% and exports, given the less favourable conditions on the international markets, should fall by around 3%. Maintaining economic growth in 2000 at a level comparable to that of 1999 could mean an increase in steel consumption of at least 5.5%, but crude steel production may well remain at the 1999 level, which would result in an increase in imports of around 4.3%, while steel exports would continue to fall by just over 5%.

The **Norwegian economy** continued to grow vigorously in 1998, as it has every year since 1993, despite a slight slowdown due to the implementation of more restrictive policies and the fall in the price of oil. Growth in GDP over the year as a whole amounted to 2.3%, while investment increased by only 5%, investment in house-building, for its part, having fallen. Household consumption also declined. The high rise in interest rates reduced activity in the construction sector and caused vehicle sales to fall off. The shipbuilding and off-shore oil sectors experienced a marked slowdown in activity towards the end of the year. According to the most recent forecasts, the global economic growth rate should slow down further in 1999 to around 2%, which could prompt a decline in investment of over 7%. In 2000, following an improvement in the world economy and a relaxation in monetary policy, economic growth can be expected to increase by around 2.6%.

Crude steel production rose by 11.4% in 1998 and production plants operated at 86% of their capacity. Apparent steel consumption rose by 11% and net steel imports by 10.8%. A slowdown in activity in the construction, shipbuilding and offshore oil sectors in 1999 should cause steel consumption to fall by more than 10%. Crude steel production may fall slightly by 6.3%. Since exports, chiefly to the EU(15), are expected to fall by around 16.7%, steel imports should fall by around 14.5%. In 2000, developments are expected to be similar to those of 1999, with steel consumption continuing its downward trend and possibly falling by 10%, steel output remaining at its 1999 level, like steel exports, and steel imports falling by nearly 10%.

Poland's economy continued to grow for the seventh year running, though at a slower rate. GDP grew by 5% in 1998 and disinflation proceeded at a steady and regular pace. Despite the slowdown in exports to Russia towards the end of the year due to adverse external factors, the economy continued to grow in 1998, as a result of faster growth in value added in the industrial sector and very strong growth in activity in the construction sector, which expanded by a further 11.6% in 1998. Private consumption continued to grow, but less rapidly – at a rate of 4.5% in 1998 – and investment rose by a further 22%. Total industrial output grew by 4.8%, with a marked slowdown during the last quarter of the year. Activity in the main steel-consuming sectors was more mixed than in previous years. Very high increases in activity were reported: around 13.9% in the automobile sector, 25.9% in the railway sector, and 40.6% in the rolling stock sector. On the other, hand output in machinery and equipment and in industrial vehicles showed a downward trend.

Despite a slowdown in production over the first quarter of 1999, macroeconomic results should remain good throughout the year with growth in GDP close to 4.5% again. Investment is therefore estimated to grow by a further 9.4% in 1999 and by 9.7% in 2000. GDP should grow by around 4.5% in 1999 and 5% in 2000. Private consumption will probably remain buoyant but should gradually taper off. The rate of unemployment should continue to fall, although at a moderate pace, and inflation should be brought down to 8.6% in 1999. Growth in industrial production should continue, though at a much lower level than in 1999, owing to export problems, but the situation should recover in 2000.

The sound performance of the Polish economy in 1998 led to an increase in steel consumption of around 5.1%. Crude steel production declined by 14.4% as a result of a fall in steel exports of 23.6% – amounting to a substantial decrease of nearly a million tonnes – and a rise in steel imports of nearly 34%. Significant falls in production affected the two main producers, namely Huta Katowice and Huta Sendzimira. With the upward trend in prices continuing throughout 1998, companies had good results despite the lower level of production. 1999 should see steel demand continue to rise, although at the lower rate of 4%. Crude steel production should continue to decline by 1.9%, but imports are likely to fall by over 9% and exports will probably decline by nearly 20%. In 2000, the positive trend in steel demand should continue, with a further increase of around 5%, while steel production should rise by more than 9%.

In **Switzerland,** steady domestic demand in 1998 enabled the economy to grow by 2.1%, despite a slowdown in exports over the second half of the year. This was the best result for the Swiss economy in the past eight years. Strong investment in machinery and plant, and faster growth in household consumption have helped fuel domestic demand. In spite of the continuing crisis in the construction sector, gross fixed capital formation increased by 3.8%. In 1999, economic activity should revert to its potential level and GDP is expected to rise by 1.5%. Activity in the construction sector should pick up for the first time in four years. In the machinery and instruments sector, growth in activity looks set to slow down, affecting steel consumption in particular. In 2000, growth should gradually rise to 1.8%, accompanied by an increase in household consumption, while investment should begin to recover and subsequently rise by 4%.

Matching the trend in the economy, steel demand rose by 1.6% in 1998 to 1.92 million tonnes in finished product equivalent. Crude steel production was down by 2.8% on 1997 to a total of 1.02 million tonnes. This decline was partly attributable to a fall in domestic orders. With regard to foreign trade, steel imports were up 5.9% on 1997 to 1.96 million tonnes. Exports, too, rose by around 5.8% to 0.96 million tonnes. As a result, net steel imports rose by around 6.4%.

In 1999, steel demand can be expected to fall by around 1.5%. This slowdown should be accompanied by a fairly limited decline in crude steel production of around 3.9%. Against this background, there could be a slight fall in the price of steel. An upward trend in steel demand is expected in 2000, accompanied by a substantial increase in production, while trade in steel products should remain stable.

In 1998, the economy of the **Czech Republic** experienced a serious recession and the fall in activity gradually increased over the year to -2.7%. Domestic demand fell by 2.1% under the combined effects of a slowdown in household consumption due to the fall in net income and a decline in investment caused by more rigorous credit conditions. The export situation, which had been fairly healthy at the beginning of the year, deteriorated as a result of the unfavourable international climate. The unemployment rate increased further, from 4.7% in 1997 to 6.5% in 1998. Investment was down by 3.7%. The economic slowdown looks set to continue in 1999 but a recovery is expected in 2000. The GDP growth rate will probably be -0.5% in 1999, rising to 2.5% in 2000. Private consumption could begin to pick up in 1999, while private investment is not likely to recover before 2000. The situation should nevertheless start to improve in the second half of 1999.

Despite the slowdown in economic growth, industrial output increased by a further 3.4% and steel demand rose by 8.5% in 1998. Crude steel production fell by 3.7% to 6.5 million tonnes, and steel exports experienced a similar downward trend, whereas imports increased. As a result, net exports fell by nearly 32%. In 1999, steel consumption is likely to decline and could fall by 12.5%. Crude steel production should fall by the same proportion, with net trade in steel probably remaining close to the 1998 level. Steel consumption is unlikely to recover before 2000, when it could rise by 7.5% to 8%, accompanied by a strong increase in crude steel production of 17.5%, a recovery in exports and a fall in imports.

In **Turkey,** economic growth slowed down significantly throughout 1998 and the GDP growth rate, which had been 9.2% over the first quarter of the year, was no more than 0.7% over the fourth quarter, amounting to around 2.8% for the year as a whole. This slowdown in growth was caused partly by the implementation of the programme to stabilise the economy and, in the second quarter, by the impact of the crisis affecting the emerging countries. Total domestic demand fell by 0.3% and investment was down by 2.4%, while private consumption was virtually flat. Inflation soared to 80.7%. Industrial output followed the same trend, actually falling by 5.1% over the fourth quarter, and increasing by only 1.8% for the year as a whole, compared with 11.5% for 1997. Apparent steel consumption increased by 6.3%, while crude steel production fell by 0.6%. The structure of the Turkish steel production sector is such that there was a further deficit in flat products and a substantial surplus in long products. As a result, imports increased by 5.1%, whereas exports, chiefly to the Far East or the NIS, fell by 11.7% to just 6.5 million tonnes.

1999 should be a year of transition for the Turkish economy. However, if the programme to stabilise the economy continues without mishap, GDP growth could slow down to just 1.4% in 1998 and investment should stagnate, while private consumption could increase by 2%. Economic growth should pick up in 2000, with GDP up by 3.9%, investment rising by 5.2% and household consumption likely to be up by 3.8%. In spite of the unfavourable economic situation, apparent steel consumption could increase by a further 2.8% in 1999, mainly reflecting the 6.2% decline in steel exports, while crude steel production should fall by around 1.8%. 2000 should see another slight increase in steel consumption of around 1.4%, with output rising by 2.2% to its previous 1997 level of 14.2 million tonnes, and exports picking up.

Japan

The Japanese economy deteriorated further in 1998, with a GDP growth rate of -2.8%. This fall mainly reflects substantial cuts both in total fixed investment, and private investment in the manufacturing and house-building sectors. Private consumption was also down, falling by 1.1%, partly because consumers uncertainty over the outlook for the economy. With final domestic demand down by 3%, stock inventories were high and production started to fall.

The construction sector, both residential and non-residential, remained on a downward trend; while industrial output fell by 6.2% in 1998. Sales of private cars and industrial vehicles in Japan were flat, and exports slowed down; in particular exports of vehicles to South-East Asia were down, with a consequent 7.7% fall in the automobile sector. Production fell by 10.6% in the machinery and equipment sector but only by 2.2% in the electrical equipment sector. Only in the shipbuilding sector did activity remain at a high level, owing to the large number of orders pending.

Indices of activity in the steel-consuming sectors, 1995 = 100

	1996	1997	1998	1998/97 % change
Industrial production	102.3	106.0	99.4	-6.2
Production of passenger cars	102.4	111.3	102.7	-7.7
Production of commercial vehicles	106.5	111.9	95.9	-14.3
Non-electrical machinery	104.8	108.3	96.8	-10.6
Electrical machinery	108.9	114.4	111.9	-2.2
Shipbuilding	106.8	107.5	108.1	+0.6

Source: OECD, *Indicators of Industrial Activity.*

In 1999, despite clear signs of improvement, the Japanese economy will still be in difficulties. The two economic stimulous packages of April and November 1998 should begin to take effect and bolster economic activity. The effects of these measures and of lower taxes should become apparent, prompting a recovery in private demand, with investment in such sectors as house-building, vehicle sales or domestic appliances. Despite these efforts, the most recent economic projections for 1999 still indicate that GDP will be very slightly down, falling by less than 1% over the year. The main indicators are expected to show a downward trend: private consumption should decline by 0.1%, investment by 3.3% and final domestic demand by 0.9%. The same forecasts also predict that the economy should stabilise in 2000, though many uncertainties will remain.

In the steel sector, apparent consumption fell by 17.2% in 1998, decreasing by 13.9 million tonnes from the previous year. Because of a slight reduction in stocks, real steel consumption only fell by around 15.3%. Because of the problems the Japanese economy is still having to cope with in 1999, together with the adjustment of steel inventories, apparent steel consumption could continue to decline by around 1%, before rebounding in 2000 with estimated growth of about 2.8%.

31

The substantial fall in steel demand and the beginnings of a reduction in stocks caused crude steel production to fall by 10.5% in 1998, 11 million tonnes down from the 1997 level. The capacity utilisation rate was only 62.3%. In 1999, a decline in domestic demand and a probable decline in exports should cause crude steel production to fall by around 6.2%, with the volume of production down by 5.8 million tonnes to around 88 million tonnes. In 2000, depending upon whether demand recovers as expected, crude steel production may begin to rise again and, with growth at just over 2%, could well return to around 90 million tonnes.

In 1998, owing to the marked decline in the demand for steel, steel imports fell by 25.5% and the share of imports in the Japanese market went down from 8% in 1997 to 7.2% in 1998. At the same time exports of Japanese steel rose by 9.4%, to nearly 25 million tonnes, the highest level in ten years. The economic and financial crisis that gripped South-East Asia from the summer of 1997 onwards caused a sharp fall of around 12.6% in steel exports to countries in that region in 1998. Exports to Korea were down by a million tonnes, a fall of 22%, while exports to Chinese Taipei rose by 121.9%. The high rise in exports was due mainly to the exceptional growth in exports to the United States, up by 259.7%, and to other non-Asian countries.

In 1999, in light of the estimated demand for steel, imports are expected to contract by around 12% to 13%. During the first quarter of 1999, exports were down by 5.2% over the first quarter of 1998. Exports to the United States market fell by 94.7% and those to the EU declined by 45%. Conversely, exports to Asian countries began to pick up, rising by 17.6%. Steel exports for 1999 as a whole should decline by around 21.5%. In 2000, the downward trend in exports could continue, though more moderately, and imports could also continue to fall slightly.

Korea

In 1998, the Korean economy experienced the most severe recession since the War. The collapse of the banking system and a string of bankruptcies paralysed industrial activity. In accordance with the IMF aid programme and the government's wish to make essential economic reforms, the financial sectors and companies underwent fundamental restructuring in 1998. These efforts succeeded in restoring a certain degree of stability and the main macroeconomic indicators began to show signs of recovery from the second half of 1998 onwards.

In 1998, GDP fell by 5.8%. The crisis caused a fall in investment of around 38.5%, and a decline in private consumption in the region of 9.6%. The unemployment rate jumped from 2.6% in 1997 to 6.8% in 1998. Exports of goods and services soared by 13.3%, stimulated by the depreciation of the won, whereas imports fell by 22% as economic activity slowed down. In 1999, the Korean economy seems well on the way to recovery. The most recent forecasts indicate that GDP growth could increase to 4.5%, with private consumption set to rise by 2.5%, and domestic demand by 6.2%. Investment should fall by a further 3.9%. Exports should rise by a further 6% and imports should begin to pick up, and subsequently rise by 12%. Despite these brighter prospects, unemployment should continue to increase as a result of the restructuring in progress in the large companies and could reach 7.6%. The economic situation can be expected to stabilise in 2000 and GDP is expected to increase by more than 4% again. All the main economic indicators should then be moving in the right direction.

Activity in the construction sector fell by 40.8% in 1998 but could begin to pick up by about 4.3% in 1999, helped by government plans for new infrastructure projects to combat the rise in unemployment. As a result of falling domestic demand and large-scale strikes, production in the automobile sector fell by 30.7% in 1998. However, a marked upturn is expected in 1999, with production rising to around 22.8%, as the economic recovery and the measures taken by the government fuel domestic demand for motor vehicles. In 1998, the machinery and plant industry suffered from a low capacity utilisation rate and from credit levels that adversely affected industrial investment. Activity in this sector was consequently down by 22.4% in 1998, but with the recovery in investment in 1999 and the trend to increasing automation, it should begin to rise again by about 6.5% in 1999. In the shipbuilding sector, production increased by 15.9% in 1998, owing to the number of orders pending and the efforts to improve productivity and reduce costs. Activity in the sector should continue to rise by around 10% in 1999, given the large number of orders already placed.

In 1998, apparent steel consumption fell by 37.1% to some 23.8 million tonnes. The predominant factor in this decline was the extremely negative trend in demand for long products. Crude steel production fell by 6.2% to 39.9 million tonnes, down by 2.7 million tonnes on 1997 levels. Owing to a reduction in capacity in the course of the year, the capacity utilisation rate was 94%. As a result of the substantial decrease in domestic demand and the fall in the value of the won, steel imports fell by 56.9% in 1998, while at the same time exports increased by 56.3%, particularly to the more vigorous markets such as the United States and Europe.

In 1999, with the recovery in the economy, apparent steel consumption may well increase by 14.3%, amounting to 3.4 million tonnes more than in 1998. Despite the improvement in demand, crude steel production should continue to fall by 3.4% in 1999, owing to adjustments that have to be made to avoid surplus capacity. The increase in domestic demand, the rising value of the won, the falling demand in the United States and the EU, and the rise in market tensions, should lead to a 23.2% fall in steel exports, while imports should continue to fall by around 14.8%.

In 2000, steel consumption may continue to grow by around 10.5%, or by nearly 3 million tonnes more than in 1999. This increase will probably be accompanied by a fall in net steel exports of around 16.3%, and a recovery in crude steel production, which could rise by about 4.2% to slightly more than 40 million tonnes again. The capacity utilisation rate could exceed 94%.

Australia and New Zealand

The **Australian** economy continued to grow in 1998, at an even faster rate than in previous years. GDP rose by 5.1%, and inflation remained very low at around 1.7%. The economy was buoyed by private consumption, which was up 4.3% and to a lesser extent by investment, which increased by a further 5.1%. Industrial output was only up 1%, but activity in the construction sector remained firmly on course. In 1999 and 2000, the economic growth rate should revert to around the average level, with GDP increasing by 3.2% to 3.4% per year. Investment could decline slightly in 1999 but should start to pick up in 2000. Private consumption will probably increase by more than 4% over the next two years and unemployment should gradually decline over the period to no more than 7.4% in 2000.

1998 was another excellent year for the Australian steel industry. Steel production increased by 1.2% to 8.9 million tonnes, a new record level. This result was achieved in spite of the slowdown in production in November and December due to the impact of the crisis in Asia, and reflects the enlargement of the Port Kembla steelworks, which produced 5 million tonnes, and of Smorgon, whose production capacity increased by 2.2% to 9.2 million tonnes in 1998. Exports of steel products also increased, up by 4.5% to a total of 3.5 million tonnes, while imports remained stable at 1.2 million tonnes. As a result, apparent steel consumption was down slightly, falling by 0.8%. In 1998, the effects of the crisis in Asia only became apparent towards the end of the year.

In 1999, demand for steel is expected to remain very high and could match the record 1997 level of 5.9 million tonnes in finished product equivalent. Steel imports could decline slightly, falling by around 3.2%, whereas exports are likely to be hit harder and could fall by nearly 20%. Such a decline in exports would have serious adverse repercussions on crude steel production, which is likely to fall by 7.2% to just 8.3 millions de tonnes. This fall would be associated with the closure of the Newcastle works at the end of September 1999 and a reduction in production capacities of about 0.5 million tonnes, representing a 4.8% reduction.

In 2000, apparent steel consumption looks set to rise by a further 1.2% to 6 million tonnes. Exports are likely to fall by a further 28.6% to just 2 million tonnes. Crude steel production should fall again by nearly 10% to no more than 7.5 million tonnes, reflecting the fact that the restructuring of the Australian steel industry has been completed, with its total capacity reduced by a further 15% to 7.5 million tonnes. The capacity utilisation rate would consequently be 100% in 2000.

New Zealand's economy was affected more seriously by the crisis in Asia and the recession did not end until the second half of 1998. Over the year, GDP declined by 0.8%. However, economic growth is likely to reach 2.6% in 1999 and should accelerate in 2000, stimulated by new policy measures, greater competitiveness and a return of consumer confidence. With regard to the steel sector, crude steel production in 1998 remained stable at its 1997 level, while an 11% increase in exports caused a reduction in apparent steel consumption of nearly 10%. In 1999, steel consumption is expected to rise by 8%, but this rise will be accompanied by a marked reduction in exports, which will result in a decline in crude steel production of nearly 8%. In 2000, the demand for steel is expected to rise by 17%, and production should be up by more than 14%.

Brazil

Brazil's economy grew by only 0.15% in 1998, while industrial output fell by 1%. This mediocre result reflects a significant decline in activity in the second half of the year, essentially due to measures taken by the government to cushion the effects of the crisis in Russia. Although several far-reaching measures had been adopted in an attempt to overcome the budgetary deficit, and despite the steep increase in interest rates, it proved impossible to prevent large amounts of currency from flowing out of the country. Inflation did not exceed 1.7%, the lowest level in 40 years, and unemployment rose to 7.6%. In January 1999, the government decided to adopt a floating exchange rate system, which led to an 80% devaluation of the Brazilian real. Since the beginning of March, the real has regained half of its value against the dollar. This rapid recovery has completely reversed the prospects for Brazil's economic development over the current year. As a result, the most recent estimates suggest that GDP will probably fall by around 3% for the year as a whole and that the rate of inflation should remain below 10%. The Brazilian economy should begin to pick up in 2000, growing by around 2%, with inflation kept below 10%.

Apparent steel consumption would seem to have grown by 3.3% in 1998 to a new record level of 14.7 million tonnes in finished product equivalent. Net exports of steel products fell by 7%, in response to a steep increase in imports, up by 26.8%, combined with a 4.4% decline in exports, which fell below the 9 million tonne mark for the first time in ten years. Crude steel production fell by 1.5% to 25.8 million tonnes.

In 1999, the decline in economic activity should result in an 8.5% fall in apparent steel consumption. Steel imports should fall by more than 10% to around the level reported in 1997. Steel exports, suffering from the slowdown in world demand should remain close to their 1998 level. This should cause a further decline in crude steel production of around 4.5%. The improved economic climate of 2000 should allow apparent steel consumption to rise by around 4.7%. Crude steel production should also be up by 3% and imports will probably decrease further, as will exports.

Other non-OECD economies

Other Latin American countries

The region as a whole was hit by the effects of the crises in Asia and Russia in 1998 and the second half of the year saw a downturn in growth followed by a fall in production in most countries. The slowing down of capital flow, tightening of credit and falling off of trade, all helped worsen the state of the economy. At the beginning of 1999, the monetary crisis in Brazil drove the whole region towards recession. The contagious effects of the Brazilian crisis on the Mercosur member countries, particularly Argentina, arose from an imbalance in trade. In Chile, these effects could be seen in the need to keep interest rates low in order to stimulate the economy and the need to ensure the stability of the national currency. Similarly, Ecuador had to confront a serious banking crisis. In short, the economic outlook in 1999 for most countries in this area is fairly bleak, with production levels falling. However a recovery, albeit modest at times, is forecast for 2000, when a fall in the real interest rate seems likely to stimulate economic activity once more.

The economic trend has of course created a most unfavourable climate for the steel industry in Latin America and apparent steel consumption fell by 7.1% in 1998. Net imports of steel to the area as a whole were down by 22% on 1997, but this was due to a substantial rise in exports. Crude steel production fell by 3% to just 11.7 million tonnes. Demand for steel is likely to worsen in 1999, and could fall by 5.4%. As a result of this decline, combined with a fall in exports, crude steel production is likely to fall by 21.8% to around 9 millions de tonnes. The year 2000 could see a fairly strong recovery in steel consumption, and apparent consumption should rise by 7.2%. An increase in steel production, which should amount to 10.5 million tonnes, up by 13.5% on 1999, should keep imports growth down to a mere 1.3%, whereas exports could rise further by around 10%.

Africa and the Middle East

In **South Africa,** apparent steel consumption declined by 14.6% in 1998 and should continue to fall by just over 2% in 1999 before starting to rise again by around 5% in 2000. Crude steel production also declined in substantial quantities, and in 1998 steel production was down by 9.6%, falling to 7.5 million tonnes in volume terms. This downward trend is expected to increase by nearly 15% in 1999, bringing levels to their lowest in more than 25 years, before production starts to pick up again in 2000. This particularly negative trend is due to conditions on the international steel market, exports of steel from South Africa being hardest hit.

Demand for steel in the rest of the **African continent** fell by a further 2% in 1998 and would appear to amount to 4 million tonnes. Crude steel production dropped for the seventh consecutive year, and the 20.7% fall reported in 1998 brought production down to a mere 0.73 million tonnes. Steel imports now account for 90% of the steel consumed in Africa, and exports are still very low and primarily limited to trade within Africa. In 1999, the African steel market is expected to deteriorate, with another fall in apparent consumption of around 6.5%, accompanied by a further decline in production of nearly 10%. An upturn in steel demand is not expected before 2000 when recovery will be modest with levels remaining low. At the same time steel production is also likely to pick up, and imports on the continent should remain steady at 90% of apparent consumption.

In the **Middle East,** steel consumption, which had been rising for ten years, suddenly fell by 12% in 1998, down by 4 million tonnes on 1997. The trend in crude steel production was also downward, the 5.4% decline reported in 1998 amounting to a fall of 0.7 million tonnes on the previous year. The average rate of capacity utilisation was around 79%. Steel imports also fell sharply, down by more than 15% to less than 20 million tonnes. Exports too declined significantly, falling by around 10%. In 1999, apparent steel consumption should continue its downward trend and could decline by a further 10%, before beginning a fairly strong recovery of around 7.5% in 2000. Crude steel production could begin to pick up from 1999 onwards, and the recovery could accelerate significantly in 2000. Steel imports should stabilise at 20% below the record 1997 level, reflecting the setting up of new production capacities, which should account for some 7.5 million tonnes throughout the region between 1998 and 2000.

Asia

The Asian continent, excluding China, can be divided into three separate regions: India, the five main members of the ASEAN defined as ASEAN(5), namely Indonesia, Malaysia, the Philippines, Singapore and Thailand, and lastly all other Asian countries, including Chinese Taipei, Pakistan, Vietnam and North Korea, which was covered with China in the past.

In **India**, the rate of economic growth remained fairly high in 1998, at around 5.2%, but industry's contribution to GDP slowed down, increasing by just 4% compared with 5.9% in 1997. Industrial output continued to rise but at a markedly lower rate than in previous years, significantly below its potential, and in many industries production actually fell in 1998. The decline in production in the industrial sector generally was due to a slowdown in domestic demand, low exports and increased competition over imports. In different industries, it was possible to observe slower investment, effects of high credit charges, and problems related to infrastructure. Exports of goods and services fell by 3.5% while imports rose by around 7%. The rise in interest rates, on the other hand, had an adverse impact on investment, which was already at a low level. The inflation rate doubled, rising from 7.2% in 1997 to 14% in 1998. The situation should, however, improve in 1999 with an expected growth of 5.7% in GDP and an upturn in exports offsetting domestic demand, which is still slow. This turn for the better should continue in 2000.

The less favourable economic climate in 1998 led to a reduction in apparent steel consumption, which fell by 2.7%. At the same time, crude steel production followed a similar trend and fell by 2.9%, falling back below the 24 million tonne mark. Net steel imports remained at the 1997 level, reflecting the simultaneous drop in imports and exports. In 1999, with the expected recovery in economic activity, the demand for steel should increase by 3.5%. A fairly marked increase – around 5.5% – in crude steel production, which should exceed 25 million tonnes for the first time, should cause a significant drop in net steel imports. In 2000, steel consumption is likely to be up by a further 4.8% on 1999. Crude steel production could well increase by around 3.3% to exceed the 26 million tonne mark and both exports and imports of steel are likely to rise.

As for the **ASEAN(5)** members which appear to have been most affected by the financial crisis that began during the summer of 1997, their economies remained in recession throughout 1998 and, although a degree of stabilisation could be observed in certain countries during the second half of 1998, unemployment continued to soar and there was little sign of recovery in production or household expenditure. In Thailand, the first economy to be affected and where the slowdown in growth had been most marked, the economy declined by 8% in 1998, with domestic demand falling by 23.5%. Of the other economies most affected in the region, Indonesia's economy contracted by 13.7%, with domestic demand falling by 20% and inflation rising from 6.6% in 1997 to 60% in 1998. Malaysia, too, was severely affected, with GDP down by 6.7% and domestic demand by 25.9%, although inflation was kept down to about 5.3%. In the Philippines, the economy shrank by only 0.5%, as domestic demand had declined by only 8%. Although it is impossible to give a detailed account here of all the repercussions of the crisis on the economies of these countries, it is nonetheless worth noting that in 1999, although the recession is likely to continue in Indonesia, a recovery should begin in Thailand and Malaysia and growth in the Philippines could be around 2.5%. Economic growth should begin to pick up more substantially in 2000 to between 3% and 4.5%, depending on the country.

With regard to the steel sector, steel consumption in these countries, having initially declined by 8.5% in 1997, fell by a further 36.8% in 1998 to no more than 19.8 million tonnes. Crude steel production fell by 17.5% to just 8.5 million tonnes. The marked reduction in demand led to a decline of nearly 45% in net steel imports, with imports down 38.4% from 24.9 million tonnes in 1997 to 15.4 million tonnes in 1998, and exports up nearly 12% to a total of 3.2 million tonnes. While steel demand followed a downward trend in the five countries, the extent of the decline differed from one country to another, with consumption falling by 43.6% in Indonesia, 20% in Malaysia, 43.5% in the Philippines, 8.6% in Singapore and 56% in Thailand.

In 1999, when the effects of the crisis should begin to fade, apparent steel consumption is expected to fall by around 4.5% in the area as a whole. Trends should differ depending on the country; a recovery is expected in the Philippines with steel demand rising by 1.5%, whereas further decline is likely in the other four countries, with rates varying between -3.5% and -7%. Crude steel production in these countries as a whole should only fall by 1.7%. Steel exports should be down by 12.2%, while imports are likely to fall by a further 7.5%, 1.2 million tonnes down on 1998.

In 2000, the steel market in these five countries should follow an upward trend, and steel consumption could recover by around 9.7%. This recovery, varying in size, is likely to benefit all these countries. As a result, steel output could recover by around 7%. Steel imports might then begin to rise slightly and could increase by just over a million tonnes, whereas exports should continue to decline.

Trends in the economies of the **other Asian countries**, of which the most important with regard to steel are Chinese Taipei, Pakistan, North Korea and Vietnam, varied substantially in 1998. North Korea continues to face serious economic problems. Chinese Taipei was the country in the group least affected by the Asian crisis and continued to report substantial economic growth in 1998. Similarly, Pakistan's economy began to pick up at the beginning of the year. In Vietnam, economic growth slowed down, but GDP increased by 5.8% in 1998. Industrial output rose by 12.1%, and in the sectors with strong foreign direct investment, output jumped by 23.3%, whereas in sectors strongly dependent on the state, growth was significantly lower, at between 7% and 8%.

Steel consumption in these countries as a whole was down by 5.2% in 1998. It is worth noting that this reduction was due to a fall in demand for steel in Chinese Taipei of around 4.5%, 1 million tonnes less than in 1997, bearing in mind that consumption in Chinese Taipei represents 78% of the total consumption in the area. Consumption also fell in North Korea and other countries in the area but in Vietnam apparent steel consumption increased by 14.5%. Crude steel production in the area increased by 4.6% to exceed 18 million tonnes, but here again it should be noted that 92.3% of this production, or 16.9 million tonnes, were accounted for by Chinese Taipei alone. Net imports of steel to the area fell sharply, down by 19% on 1997. 1999 should see steel demand picking up by around 1%, though in Chinese Taipei the increase is likely to be in the region of 2.2%. This increase, combined with a subsequent slowdown in net imports of steel, should result in an increase of nearly 5% in crude steel production in the zone. In 2000, steel demand should start to increase more significantly, rising by nearly 10%. Steel production will probably increase only modestly, by no more than about 2%, and the increase in demand will therefore have to be met by net imports, which will probably rise more substantially, by over 27%.

Since 1996, when the Czech Republic, Hungary and Poland became Members of the OECD, the only countries left in this area have been Albania, Bulgaria, the Slovak Republic and Romania. The steel market in Albania can still be regarded as flat. For the eighth year running crude steel production in Albania has remained at 0.02 million tonnes and no significant change in the situation is foreseen over the next few years.

In **Bulgaria,** structural reform has advanced considerably since 1997, particularly in respect of privatisation and the reform of the banking sector. The severe recession in which the country was immersed for several years ended in 1998 and GDP rose by 2.7%, supported by an increase in industrial activity of around 4%. Privatisation of companies has accelerated and 31% of Bulgarian enterprises are now privately owned, with the proportion expected to rise to 50% by the end of 1999 as the pace of privatisation increases. In 1999, economic activity is expected to slow down; GDP will probably increase by only 1%, and industrial output will mark time. There is likely to be increased activity in the construction sector, however, as infrastructure projects are implemented, as well as improved output in the manufacturing industries following the steep decline reported in previous years. The economic recovery should accelerate in 2000, with growth likely to reach 3%.

Apparent steel consumption, which had begun to pick up in 1997, increased by a further 8.2% in 1998 but at 0.7 million tonnes it is still at a very low level. Crude steel production fell by 15.7%, mainly because of an 18% fall in steel exports. In 1999, steel consumption should be virtually flat, but it will probably increase more significantly from 2000, by around 31%. Steel production should fall by 11.3% in 1999 as a result of the decline in exports but it could pick up in 2000 and increase by nearly 30%.

GDP growth in the **Slovak Republic** slowed down in 1998 to just 4.5%. The build-up of a sizeable budgetary deficit, in excess of 10.5% of GDP, is putting an end to the strong growth reported in recent years. Macroeconomic stabilisation measures, together with reforms in the banking sector and in companies, will force growth downwards so that it is unlikely to increase by more than 2% in 1999 and 2000. Domestic demand slowed down and unemployment rose. As a result of the fall in the growth rate, unemployment could be worse still in 1999, peaking at 15% before beginning to fall in 2000.

Apparent steel consumption fell by 25.2% in 1998. It should pick up again in 1999, rising by around 8.4%, and increase by a further 8.9% in 2000. As a result of falling demand, maintenance work on a blast furnace, and financial problems, crude steel production fell by 10.3% to 3.4 million tonnes in 1998. It should fall again by about 10% in 1999 before rising by nearly 15% in 2000. Steel imports were up by 19.4% in 1998, but should return to their usual level of about 0.7 million tonnes in 1999 and 2000. Exports were up by 3.2% in 1998, but the unfavourable climate will probably cause them to fall by around 18% in 1999, before they begin to pick up again in 2000.

In **Romania,** economic activity in 1998 was down by 7.3% on 1997, investment fell by 18.6% and inflation rose to 59.1%. The value of the currency experienced a further fall of nearly 33% against the dollar. Investment declined by 18.6% and unemployment rose dramatically. Industrial output was substantially down, by nearly 17%, with lower production in nearly all sectors, particularly the steel-consuming ones: mining, building, metal structure manufacturing, machinery and plant, whereas activity in the vehicles sector increased. 1999 could be again a difficult year for the Romanian economy, with GDP possibly falling to 4%, investment declining by 8% and industrial output falling by 6%. The economic upturn should begin in 2000, with GDP rising by 2% and investment by 5%.

As a result of the poor economic climate, apparent steel consumption was flat in 1998. Steel consumption is likely to fall by 11% in 1999 and begin to pick up in 2000, rising by around 4.8%. Crude steel production fell by 5.6% in 1998, and should decline by a further 9.7% in 1999, before rising again by just over 10% in 2000. Steel imports rose by 11.6% in 1998 but should fall quite sharply in 1999 and 2000. Steel exports, down by 6.2% in 1998, should fall by a further 8.3% in 1999 and then begin to pick up, rising by nearly 15% in 2000.

New Independent States

After a slight economic recovery beginning in Russia in 1997, the financial crisis that broke in August 1998 produced a further 4% reduction in GDP, accompanied by a sharp decline in investment and industrial output and a dramatic increase in inflation. Economic growth should contract by a further 1% in 1999, despite clear signs of recovery in industrial output. The economic recovery should not become apparent until the beginning of 2000, with GDP set to grow by around 2%. The effects of the crisis in Russia were felt by most of the other NIS in 1998 and will persist in 1999. In the Baltic States, the loss of commercial outlets in Russia accordingly caused a marked slowdown in economic growth.

Despite progress made in stabilising its economy, The Ukraine experienced a further 1.7% fall in GDP as a result of the 1998 crisis, notwithstanding the fact that 1998 had promised to be the first year of positive growth since its independence. Industrial output continued to decline, falling by a further 1.5% in 1998. Accumulated losses in industrial output between 1990 and 1998 amounted to 58%. Inflation, which had been more or less under control in 1997, rose once again, with the hryvnya rapidly losing value, and reached 20% in 1998. Domestic demand was also depressed by a fall in disposable income of more than 10%. Unemployment rose to 3.7%. The pace of structural reform, particularly privatisation, remained sluggish. In 1999, The Ukraine is likely to experience a further contraction of its GDP of up to 3.2% according to estimates. Industrial output, too, could fall by 3.5%. The economy is not likely to stabilise until 2000, with growth following a slightly more positive trend and industrial output finally starting to pick up, albeit modestly. In Kazakhstan, 1998 was marked by a reduction in GDP of about 2.5% and a fall in industrial output of 2.1%. The economy is likely to remain in recession throughout 1999 and GDP should contract further by about 2%, but 2000 should see a significant recovery, with a growth in GDP of around 3%.

In the NIS area as a whole, apparent steel consumption declined in 1998, and was 8.7% down on the 1997 level, falling below the 22 million tonne mark. It would appear to have fallen by 4.4% in Russia, 7% in The Ukraine and as much as 32.9% in the other NIS. Crude steel production declined by 8.1% to just 74.4 million tonnes. The fall in production was due to the considerable reduction in steel exports. Output fell by 9.6% in Russia, 4.6% in The Ukraine and 10.5% in the other NIS. Steel imports to the NIS as a whole fell by 31.8% in 1998, while exports, hit by declining demand in Asia, were down by 11%.

In 1999, total apparent steel consumption in the NIS should increase by 17.6%, and may well reach a total of 26 million tonnes, rising by 6.4% in Russia, 3% in The Ukraine and 14.4% in the other states. Imports may well decline slightly or remain flat but exports are likely to continue their downward trend over the second half of 1998 and at the beginning of 1999, and could fall by a further 6.5% to 2.5 million tonnes less than in 1998 as a result of lower world demand and trading problems. Crude steel production in the NIS should be up slightly, rising by about 2.2%. In 2000, steel demand should increase by nearly 9.5% in the NIS as a whole, and apparent consumption is likely to exceed 28 million tonnes, its highest level since 1994, though a mere 22% of its 1987 record. Crude steel production should only increase by around 3%, and exports are likely to fall by a further 3%.

Capacities and crude steel production in 1998

	Capacity	Production	Utilisation rate
	In thousands of tonnes		%
Belarus	1 200	1 299	108.3
Kazakhstan	6 300	3 089	49.0
Federation of Russia	91 600	43 796	47.8
Ukraine	60 730	24 461	40.3
Uzbekistan	1 170	343	29.3
Total CIS	161 000	72 988	45.3
Azerbaijan	800	250	31.3
Moldova	800	718	89.8
Georgia	1 400	444	31.7
Total NIS	164 000	74 400	45.4

Source: OECD, *Developments in Steelmaking Capacity in Non-OECD Countries, 1999 Edition.*

China

In China, growth remained buoyant in 1998, even increasing over the second half, as a result of higher public investment in infrastructure and greater flexibility in monetary policy, and GDP grew by 7.8% over the year, falling just short of the 8% goal set by the government. Domestic demand increased by 7.7% and inflation fell by 2.6%. Exports increased again in volume terms, but because of the fall in prices, their value remained the same. Interest rates were lowered in October to stimulate growth in consumption and encourage investment. The Chinese economy will probably continue its soft landing in 1999 and 2000, and GDP should continue to grow by around 8% over the first half of 1999 before slowing down slightly over the second half, so that growth for the year as a whole will be in the region of 7.2%. In 2000, growth in the Chinese economy should still be around 6.8%, but this will depend far more on trends in private consumption and company investment than on government programmes.

In 1998, steel consumption rose by a further 8.4%, or 8.3 million tonnes, to a total 105.8 million tonnes. Crude steel production was up by a further 5.5 million tonnes, an increase of 5%, to a new record level of 114.4 million tonnes. China remained the world's largest steel producer. Steel imports declined by 3.7% while exports, affected by the crisis in the Asia region, fell by 34.7% to just 5.7 million tonnes.

In 1999, as a result of the restructuring and mergers now underway in the Chinese steel industry, crude steel production will probably decline by around 3.6% to 110 million tonnes. Exports are expected to fall by around 21%, as a result of a general reduction in demand. Imports of products with high value added, however, may well remain at their 1998 level. Apparent steel consumption should therefore decline by 2% in 1999.

In 2000, renewed growth of around 5% in crude steel production will probably boost the volume of production, pushing it back up over the 115 million tonne mark. In the event of a general recovery in steel demand, referred to above, exports might well rise, albeit fairly modestly, causing a decline in imports, with steel consumption picking up by around 2%.

TRENDS IN EMPLOYMENT IN THE STEEL INDUSTRY IN OECD MEMBER COUNTRIES

In 1998, the number of people employed in the steel industry in the OECD area fell by 17 500, a decline of around 2%. Since 1974, the total number of jobs in the steel industry in the OECD area has fallen by 60.2%.

With regard to employment in the steel industry in the European Union, efforts to streamline and improve the competitiveness of the industry are continuing. In 1998, the reduction in the workforce in the EU(15) as a whole amounted to around 1.4%, a loss of 4 100 jobs. This downward trend in employment looks set to continue for several years to come.

In other European countries, although employment appears to have remained stable in Turkey, the number of workers in Norway fell by 39.2%. The total number of workers employed in the steel industry in Switzerland was down by 27.3%. The number of workers fell from 973 in 1997 to 707 in 1998. The number employed directly in production declined by 64.8%, falling from 558 in 1997 to 196 in 1998. The decline is due to restructuring.

Employment in the Japanese steel industry fell by 5.8% in 1998, amounting to a loss of some 8 500 jobs. Employment in the steel industry in Korea would appear to have declined by 9.6%, a loss of around 6 300 jobs.

Employment in the steel industry in Canada increased by 3.2% in 1998, bringing the total number employed to 34 484. In 1999, the level of employment in the steel industry should remain close to the 1998 level.

In the United States, the downward trend in employment in the steel industry gathered pace in 1998. Total employment fell by 1.4% to 232 800 workers. In 1999, the employment situation in the US steel industry could deteriorate more quickly, bearing in mind that more than 11 000 jobs were lost between January 1998 and January 1999.

Graph 4. Developments in crude steel capacity and production in the main areas of the world
Million tonnes

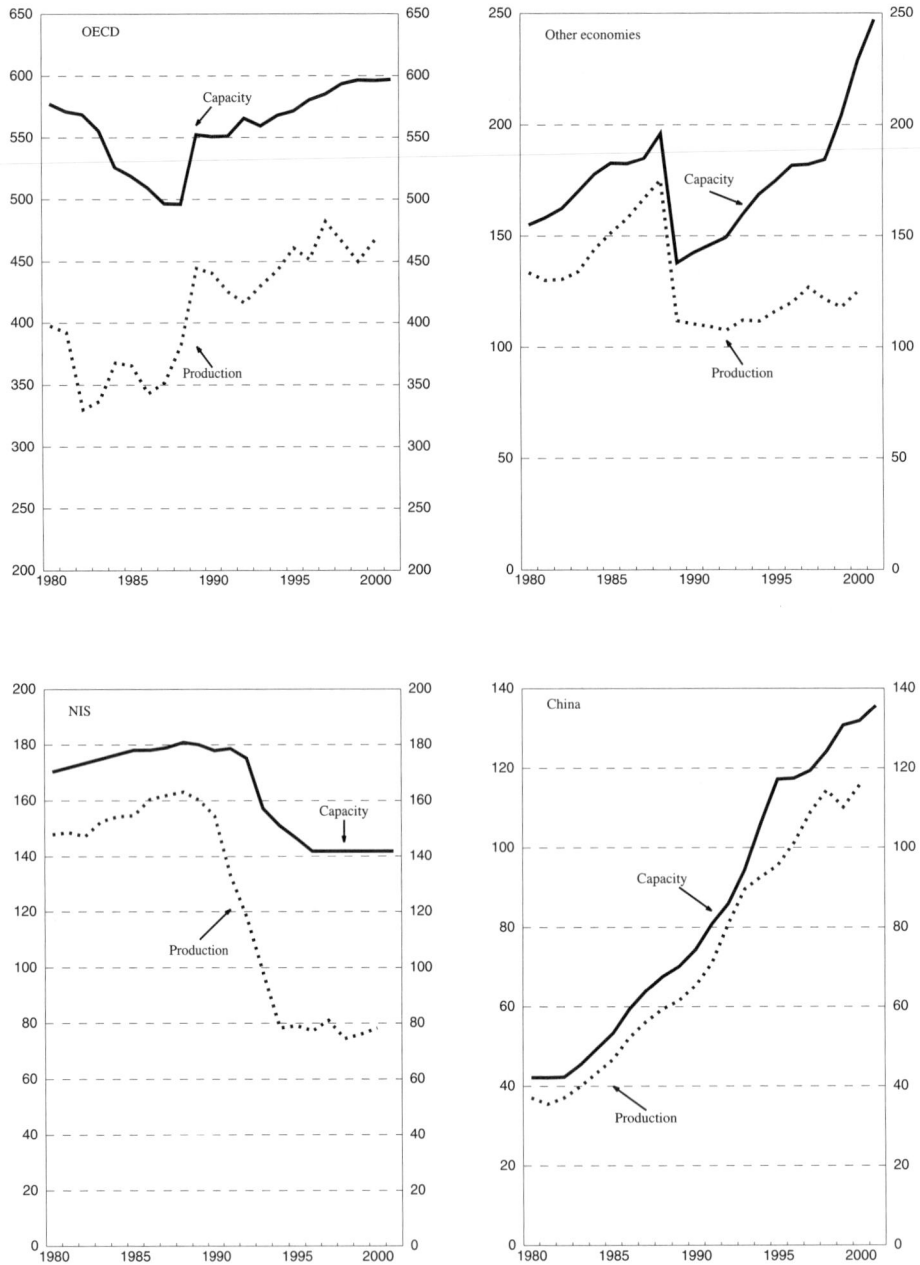

OECD

Capacity

Production

Other economies

Capacity

Production

NIS

Capacity

Production

China

Capacity

Production

Source: OECD.

Graph 5. **Developments in steel consumption and production in the main areas of the world**
Million tonnes of finished product equivalent

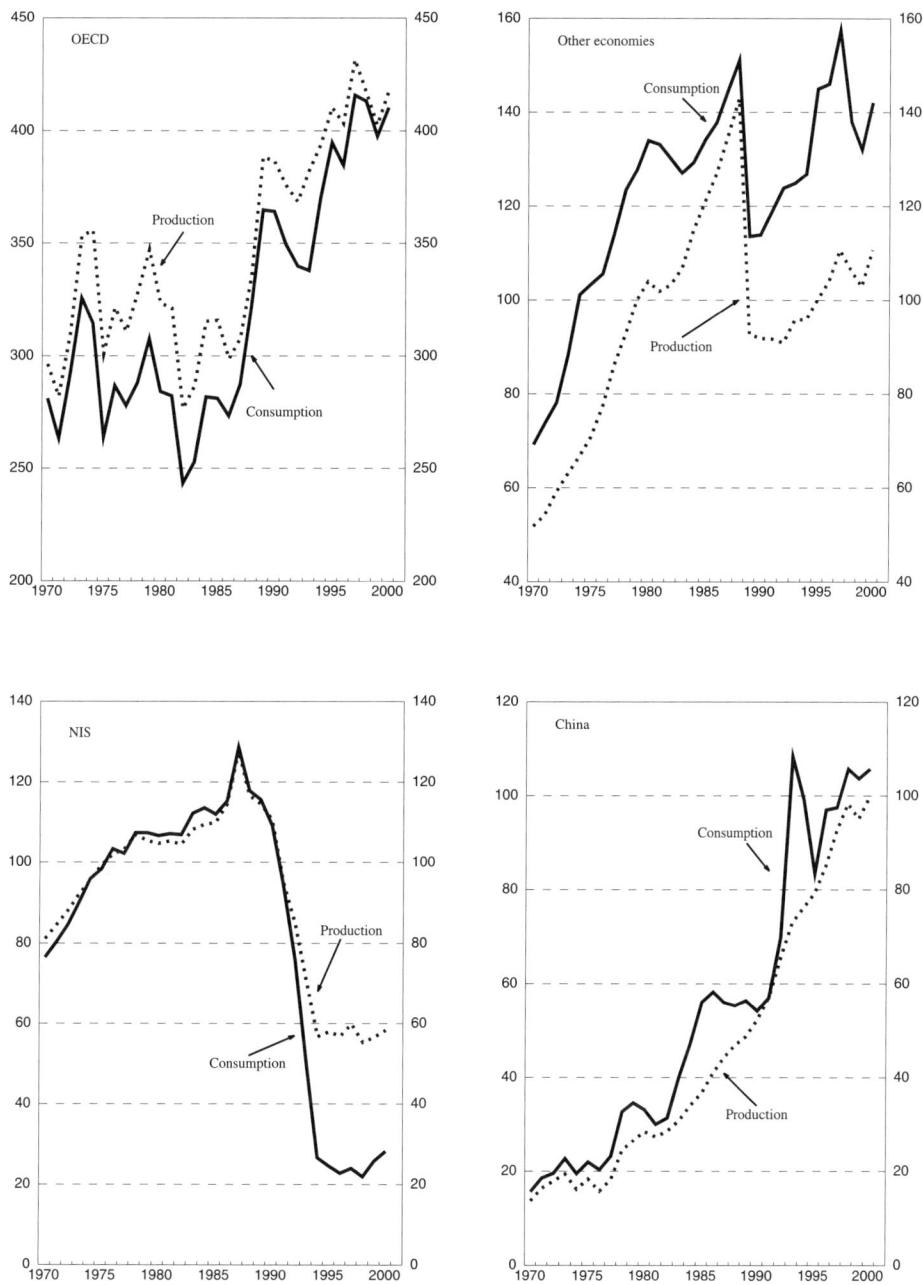

OECD

Production

Consumption

Other economies

Consumption

Production

NIS

Production

Consumption

China

Consumption

Production

Source: OECD.

STATISTICAL ANNEX

In order to reflect the expansion of the Organisation which occurred in 1996, data on the Czech Republic, Hungary, Korea and Poland have been included, wherever possible, in the "OECD Total", and to make the tables consistent, the series have been recalculated on this basis for past years.

The main changes already introduced last year compared with previous reports are the following:

-   European Union (15) has replaced EU(12).

-   The "other Europe" area covers the following countries: Norway, Switzerland, Turkey, the Czech Republic, Hungary and Poland, the latter three countries having been removed from the "Central and Eastern Europe" zone.

-   The "Central and Eastern Europe" zone now covers Bulgaria, Romania and the Slovak Republic, as well as (where possible) Albania.

-   Brazil, a full participant in the Steel Committee, now appears in almost all tables and has been excluded from the "other Latin America" area.

-   "Other Asia" has now been split in two areas: "ASEAN(5)", which groups Indonesia, Malaysia, the Philippines, Singapore and Thailand, and a new "other Asia", which groups the remaining countries, including North Korea which is no longer included with China.

-   Wherever possible, in addition to data for the New Independent States (NIS) (the former Soviet Union), the Secretariat has provided a breakdown for Russia, the Ukraine and the other NIS.

Table 1. **Apparent steel consumption** (million tonnes of product equivalent)

Tableau 1. **Consommation apparente d'acier** (en millions de tonnes d'équivalent produits)

	1990	1994	1995	1996	1997	1998	1999	2000	19/97 in/en %	19/98 in/en %	2000/99	
United States	86.0	101.6	97.5	104.2	108.3	117.0	107.1	110.3	+8.1	-8.5	+3.0	États-Unis
Canada	9.3	12.7	13.1	12.7	15.4	16.9	16.3	17.0	+9.3	-3.5	+4.1	Canada
EU(15)	119.3	113.6	132.6	115.5	131.9	143.1	135.8	138.0	+8.5	-5.1	+1.6	UE(15)
Other Europe*	25.0	22.1	25.3	23.8	27.1	29.1	28.9	29.8	+7.1	-0.5	+3.3	Autres Europe*
Japan	92.9	74.9	79.6	78.7	81.0	67.1	66.4	68.3	-17.2	-1.0	+2.8	Japon
Australia and New Zealand	5.3	6.0	6.3	6.1	6.4	6.4	6.4	6.6	0.0	-0.2	+2.6	Australie et Nouvelle-Zélande
Mexico	6.7	8.7	4.7	6.4	7.6	9.2	9.5	10.1	+22.0	+2.8	+7.0	Mexique
Korea	19.6	30.3	35.6	37.3	37.8	23.8	27.2	30.2	-37.1	+14.3	+10.8	Corée
OECD	364.0	369.9	394.7	384.6	415.6	410.6	397.5	410.2	-0.6	-3.8	+3.2	OCDE
Brazil	8.6	11.0	12.1	12.0	14.3	14.7	13.5	14.1	+3.3	-8.5	+4.7	Brésil
OECD Steel Committee	372.6	380.9	406.8	396.6	429.9	427.4	411.0	424.3	-0.6	-3.9	+3.2	Comité de l'acier de l'OCDE
Other Latin America	6.9	11.3	12.2	12.0	13.2	12.3	11.6	12.4	-7.1	-5.4	+7.2	Autres Amérique latine
South Africa	4.7	3.8	4.7	3.7	5.0	4.3	4.2	4.4	-14.6	-2.1	+5.0	Afrique du Sud
Other Africa	4.8	4.7	3.9	4.4	4.1	4.0	3.8	4.0	-2.0	-6.5	+5.3	Autres Afrique
Middle East	17.4	23.2	28.0	29.0	34.7	30.5	27.1	29.2	-12.0	-11.2	+7.6	Moyen-Orient
India	12.9	16.8	19.1	20.5	21.1	20.5	21.2	22.2	-2.7	+3.5	+4.8	Inde
ASEAN(5)**	17.3	26.8	33.3	34.2	31.3	19.8	18.9	20.7	-36.8	-4.5	+9.7	ASEAN(5)**
Other Asia	22.4	24.0	27.0	25.0	28.2	26.8	27.1	29.7	-5.2	+1.0	+9.9	Autres Asie
Total	86.4	110.6	128.2	128.8	137.6	118.2	113.9	122.6	-14.1	-3.6	+7.6	Total
Central and Eastern Europe	11.9	4.1	4.3	4.6	5.0	4.8	4.4	4.9	-4.4	-6.9	+9.7	Europe centrale et orientale
Of which :												dont :
Romania	6.3	2.6	2.8	3.1	3.3	3.3	2.9	3.1	0.0	-11.0	+4.8	Roumanie
Slovak Republic	4.3	0.6	0.6	1.0	1.1	0.8	0.9	1.0	-25.2	+8.4	+8.9	République slovaque
NIS Total	109.2	26.6	24.6	22.7	24.0	21.9	25.8	28.3	-8.7	+17.6	+9.5	Total NEI
Russia	..	14.1	15.8	13.4	13.7	13.1	14.0	15.3	-4.4	+6.4	+9.4	Russie
Ukraine	..	8.7	5.9	6.3	7.4	6.9	7.1	8.1	-7.0	+3.0	+13.6	Ukraine
Other NIS	..	3.8	2.9	3.0	2.9	1.9	4.7	4.9	-32.9	+14.4	+3.8	Autres NEI
China	54.3	99.2	83.4	97.0	97.5	105.8	103.7	105.8	+8.4	-2.0	+2.0	Chine
World	634.4	622.5	647.7	650.4	694.5	678.7	659.0	686.2	-2.3	-2.9	+4.1	Monde

Source: OECD/OCDE.

47

Table 2. **Trade balance (- = net exports, + = net imports)** (million tonnes)
Tableau 2. **Balance des échanges (- = exp. nettes, + = imp. nettes)** (en millions de tonnes)

	1990	1993	1994	1995	1996	1997	1998	1999	2000	
EU(15)	-13.0	-26.8	-22.3	-6.9	-16.0	-11.7	-0.5	-5.3	-7.0	UE(15)
Japan	-9.5	-16.6	-16.7	-15.0	-13.3	-16.3	-20.1	-15.4	-15.2	Japon
Total	-22.5	-43.4	-39.0	-21.9	-29.3	-28.0	-20.6	-20.7	-22.2	Total
United States	11.7	14.1	23.8	15.7	21.9	23.2	33.3	27.0	24.8	États-Unis
Canada	-1.0	-1.5	0.8	0.7	-0.1	2.1	3.2	2.4	3.1	Canada
Korea	-1.9	-5.7	-1.2	1.3	1.0	-1.9	-13.5	-8.8	-7.4	Corée
Other Europe	-7.3	-5.3	-6.2	-4.1	-5.4	-4.6	-1.4	-0.2	-1.5	Autres Europe
Australia and New Zealand	-1.2	-2.2	-2.4	-2.1	-2.2	-2.2	-2.3	-1.7	-0.9	Australie et Nouvelle-Zélande
Mexico	-0.4	-0.4	0.3	-5.3	-4.7	-4.4	-2.7	-3.2	-3.2	Mexique
Total	-0.1	-1.0	15.1	6.2	10.7	12.2	16.6	15.5	14.9	Total
OECD	-22.6	-44.3	-23.7	-15.6	-18.6	-15.8	-4.0	-5.1	-7.3	OCDE
Brazil	-8.8	-12.0	-10.9	-9.4	-9.9	-8.5	-7.9	-8.1	-8.1	Brésil
OECD Steel Committee	-31.4	-56.3	-34.6	-25.0	-28.5	-24.3	-11.9	-13.2	-15.4	Comité de l'acier de l'OCDE
Other Latin America	-0.9	2.1	2.5	3.0	1.8	2.7	2.1	3.7	3.5	Autres Amérique latine
South Africa	-2.8	-3.5	-3.9	-3.1	-3.4	-2.5	-2.5	-1.6	-1.6	Afrique du Sud
Other Africa	3.2	3.6	3.5	2.8	3.5	3.3	3.4	3.2	3.3	Autres Afrique
Middle East	11.2	13.0	13.1	17.4	17.6	22.5	19.0	15.1	16.0	Moyen-Orient
India	1.1	0.1	0.8	0.9	0.7	0.6	0.5	0.2	0.5	Inde
ASEAN(5)	12.2	16.6	19.8	24.1	24.4	22.1	12.2	11.4	12.7	ASEAN(5)
Other Asia	7.0	14.6	11.4	15.2	12.3	12.1	9.8	9.2	11.7	Autres Asie
Total	31.0	46.5	47.2	60.3	56.9	60.8	44.5	41.2	48.1	Total
Central and Eastern Europe	-1.8	-5.6	-5.5	-6.3	-5.2	-5.5	-4.9	-4.3	-5.1	Europe centrale et orientale
Romania	-1.3	-2.2	-2.0	-2.4	-1.7	-2.0	-1.8	-1.7	-2.0	Roumanie
Slovak Republic	-0.2	-2.5	-2.6	-2.7	-2.1	-2.1	-2.1	-1.7	-2.0	République slovaque
NIS	-1.4	-20.3	-30.1	-33.3	-34.1	-35.9	-33.4	-30.8	-30.0	NEI
China	2.0	35.4	23.0	4.1	11.9	4.9	7.4	8.5	5.9	Chine
Unspecified	0.1	0.0	0.2	0.0	1.0	0.0	1.7	1.5	1.4	Non spécifié

Source: OECD/OCDE.

Table 3. **Crude steel production** (million tonnes)
Tableau 3. **Production d'acier brut** (en millions de tonnes)

	1990	1994	1995	1996	1997	1998	1999	2000	98/97 in/en %	99/98 in/en %	2000/99	
United States	89.7	91.2	95.2	95.5	98.5	97.7	92.8	99.1	-0.9	-5.0	+6.8	États-Unis
Canada	12.3	13.9	14.4	14.6	15.5	15.8	16.1	16.1	+2.5	+1.7	0.0	Canada
EU(15)	148.4	151.7	155.8	146.6	159.8	159.6	156.8	161.2	-0.1	-1.8	+2.8	UE(15)
Other Europe	40.7	34.7	35.7	34.7	37.4	35.6	33.9	36.5	-4.8	-4.9	+7.7	Autres Europe
Japan	110.3	98.3	101.6	98.8	104.6	93.6	87.8	89.6	-10.5	-6.2	+2.1	Japon
Australia and New Zealand	7.4	9.2	9.3	9.2	9.6	9.7	9.0	8.3	+1.1	-7.2	-7.8	Australie et Nouvelle-Zélande
Korea	23.1	33.8	36.8	38.9	42.6	39.9	38.6	40.2	-6.2	-3.4	+4.2	Corée
Mexico	8.7	10.3	12.2	13.2	14.3	14.2	15.1	16.0	-0.3	+6.3	+6.0	Mexique
OECD	440.7	443.0	460.9	451.5	482.1	466.1	449.9	466.8	-3.3	-3.5	+3.8	OCDE
Brazil	20.6	25.7	25.1	25.2	26.2	25.8	24.6	25.3	-1.5	-4.5	+2.9	Brésil
OECD Steel Committee	461.3	468.7	486.0	476.7	508.3	491.9	474.5	492.1	-3.2	-3.5	+3.7	Comité de l'acier de l'OCDE
Other Latin America	9.2	10.1	10.5	11.6	12.0	11.7	9.1	10.4	-3.0	-21.8	+13.5	Autres Amérique latine
South Africa	8.6	8.5	8.7	8.0	8.3	7.5	6.4	6.7	-9.6	-14.8	+3.9	Afrique du Sud
Other Africa	2.0	1.5	1.4	1.2	0.9	0.7	0.7	0.8	-20.7	-9.6	+18.2	Autres Afrique
Middle East	6.8	11.3	11.7	12.7	13.5	12.8	13.3	14.7	-5.4	+3.9	+10.1	Moyen-Orient
India	15.0	19.3	22.0	23.8	24.6	23.9	25.2	26.0	-2.9	+5.5	+3.3	Inde
ASEAN(5)	5.8	7.7	10.2	10.9	10.2	8.5	8.3	8.9	-17.5	-1.7	+7.2	ASEAN(5)
Other Asia	18.0	13.8	12.3	13.9	17.5	18.3	19.2	19.5	+4.6	+4.9	+1.3	Autres Asie
Total	65.4	72.2	77.3	82.1	87.0	83.4	82.2	87.0	-4.1	-1.4	+5.8	Total
Central and Eastern Europe	17.4	12.2	13.2	12.2	13.1	12.0	10.8	12.4	-8.3	-10.1	+14.9	Europe centrale et orientale
Of which:												*Dont :*
Romania	9.8	5.8	6.6	6.1	6.7	6.4	5.8	6.4	-5.6	-9.7	+10.2	Roumanie
Slovak Republic	5.5	4.0	3.9	3.6	3.8	3.4	3.1	3.5	-10.3	-10.0	+14.8	République slovaque
NIS	154.4	78.3	79.1	77.2	81.0	74.4	76.1	78.2	-8.1	+2.2	+2.9	NEI
Of which:												*Dont :*
Russia	..	48.8	51.6	49.3	48.5	43.8	44.7	45.9	-9.6	+2.0	+2.7	Russie
Ukraine	..	24.1	22.3	22.3	25.6	24.5	25.1	25.2	-4.6	+2.5	+0.4	Ukraine
Other NIS	..	5.4	5.2	5.6	6.9	6.2	6.3	7.2	-10.5	+2.3	+14.0	Autres NEI
China	65.4	92.6	95.4	101.2	108.9	114.4	110.2	115.7	+5.0	-3.6	+5.0	Chine
World	763.9	725.5	751.5	749.9	799.0	776.2	754.1	785.8	-2.8	-2.9	+4.2	Monde

Source: OECD/OCDE.

Table 4. **Steel production, consumption and trade** (million tonnes)
Tableau 4. **Production, consommation et échanges d'acier** (en millions de tonnes)

1997	Production Crude steel/ Acier brut	Production via c.c.	Product eq./ Équiv. produits	Imports/ Importations	Exports/ Exportations	Balance/ Solde	Apparent consumption/ Consommation apparente	1997
United States	98.5	93.3	85.1	28.9	5.6	23.3	108.3	États-Unis
Canada	15.5	15.2	13.3	6.3	4.2	2.1	15.4	Canada
EU(15)	159.8	151.8	143.6	16.4	28.1	-11.7	131.9	UE(15)
Other Europe	37.4	28.0	31.7	14.8	19.4	-4.6	27.1	Autres Europe
Japan	104.6	100.8	97.4	6.5	22.8	-16.3	81.0	Japon
Australia and New Zealand	9.6	9.5	8.7	1.5	3.8	-2.2	6.4	Australie et Nouvelle-Zélande
Korea	42.6	41.6	39.7	9.4	11.3	-1.9	37.8	Corée
Mexico	14.3	12.1	12.0	1.6	6.0	-4.4	7.6	Mexique
OECD	482.2	452.3	431.4	85.4	101.1	-15.8	415.6	OCDE
Brazil	26.2	19.3	22.7	0.7	9.2	-8.5	14.3	Brésil
OECD Steel Committee	508.3	471.6	454.1	86.1	110.3	-24.3	429.9	Comité de l'acier de l'OCDE
Other Latin America	12.0	10.6	10.5	6.4	3.7	2.7	13.2	Autres amérique latine
South Africa	8.3	8.1	7.5	0.3	2.8	-2.5	5.0	Afrique du Sud
Other Africa	0.9	0.4	0.8	3.5	0.2	3.3	4.1	Autres Afrique
Middle East	13.5	13.3	12.2	23.5	1.0	22.5	34.7	Moyen-Orient
India	24.6	11.9	20.5	2.0	1.5	0.6	21.1	Inde
ASEAN(5)	10.2	10.2	9.3	24.9	2.9	22.1	31.3	ASEAN(5)
Other Asia	17.5	16.3	16.2	16.6	4.5	12.1	28.2	Autres Asie
Total	87.0	70.8	77.0	77.2	16.6	60.8	137.6	Total
Central and Eastern Europe	13.1	7.5	10.5	1.3	6.9	-5.5	5.0	Europe centrale et orientale
Of which:								Don't :
Romania	6.7	3.1	5.3	0.4	2.4	-2.0	3.3	Roumanie
Slovak Republic	3.8	3.8	3.2	0.7	2.9	-2.1	1.1	République Slovaque
NIS	81.0	30.3	59.9	6.8	42.7	-35.9	24.0	NEI
Of which:								Don't :
Russia	48.5	22.7	36.4	3.4	26.1	-22.7	13.7	Russie
Ukraine	25.6	5.1	18.4	1.0	12.0	-11.0	7.4	Ukraine
Other NIS	6.9	2.5	5.1	2.4	4.6	-2.2	3.0	Autres NEI
China	108.9	66.1	92.7	13.6	8.8	4.9	97.5	Chine
World	799.0	646.3	694.5	185.1	185.1	0.0	694.5	Monde

Source: OECD/OCDE.

50

Table 5. **Steel production, consumption and trade** (million tonnes)

Tableau 5. **Production, consommation et échanges d'acier** (en millions de tonnes)

1998	Production			Imports/ Importations	Exports/ Exportations	Balance/ Solde	Apparent consumption/ Consommation apparente	1998
	Crude steel/ Acier brut	Via c.c.	Product eq./ Équiv. produits					
United States	97.7	92.1	84.3	38.3	5.0	33.3	117.6	États-Unis
Canada	15.8	15.6	13.6	7.4	4.2	3.2	16.9	Canada
EU(15)	159.6	153.2	143.9	23.4	24.0	-30.5	143.1	UE(15)
Other Europe	35.6	28.2	30.4	15.8	17.2	-1.4	29.1	Autres Europe
Japan	93.6	90.7	87.2	4.8	24.9	-20.1	67.1	Japon
Australia and New Zealand	9.7	9.7	8.8	1.5	3.9	-2.3	6.4	Australie et Nouvelle-Zélande
Korea	39.9	39.3	37.3	3.6	17.0	-13.5	23.8	Corée
Mexico	14.2	12.1	11.9	2.4	6.5	-2.7	9.2	Mexique
OECD	466.1	441.0	417.1	97.1	101.1	-4.0	413.1	OCDE
Brazil	25.8	19.0	22.4	0.9	8.8	-7.9	14.5	Brésil
OECD Steel Committee	491.9	460.0	439.5	98.0	109.9	-11.9	427.6	Comité de l'acier de l'OCDE
Other Latin America	11.7	10.3	10.2	6.9	4.8	2.1	12.3	Autres Amérique latine
South Africa	7.5	7.4	6.8	0.3	2.8	-2.5	4.3	Afrique du Sud
Other Africa	0.7	0.4	0.6	3.6	0.2	3.4	4.0	Autres Afrique
Middle East	12.8	12.6	11.5	19.9	0.9	19.0	30.5	Moyen-Orient
India	23.9	11.9	20.0	1.4	0.8	0.5	20.5	Inde
ASEAN(5)	8.5	8.5	7.6	15.4	3.2	12.2	19.8	ASEAN(5)
Other Asia	18.3	17.4	17.0	15.7	5.9	9.8	26.8	Autres Asie
Total	83.4	68.5	73.7	63.2	18.6	44.5	118.2	Total
Central and Eastern Europe	12.0	7.1	9.7	1.6	6.5	-4.9	4.8	Europe centrale et orientale
Of which:								dont :
Romania	6.4	3.2	5.1	0.5	2.3	-1.8	3.3	Roumanie
Slovak Republic	3.4	3.4	2.9	0.9	2.9	-2.1	0.8	République slovaque
NIS	74.4	29.9	55.3	4.6	38.0	-33.4	21.9	NEI
Of which:								dont :
Russia	43.8	22.7	33.2	2.9	23.0	-20.1	13.1	Russie
Ukraine	24.5	4.3	17.5	0.9	11.5	-10.6	6.9	Ukraine
Other NIS	6.2	3.0	4.6	0.8	3.5	-2.7	1.9	Autres NEI
China	114.4	77.3	98.4	13.1	5.7	7.4	105.8	Chine
World	776.2	644.3	676.9	180.4	178.7	1.7	678.7	Monde

Source: OECD/OCDE.

51

Table 6. **Steel production, consumption and trade** (million tonnes)
Tableau 6. **Production, consommation et échanges d'acier** (en millions de tonnes)

1999	Production			Imports/ Importations	Exports/ Exportations	Balance/ Solde	Apparent consumption/ Consommation apparente	1999
	Crude steel/ Acier brut	via c.c.	Product eq./ Équiv. produits					
United States	92.8	87.8	80.1	31.4	4.4	27.0	107.1	États-Unis
Canada	16.1	15.9	13.9	6.6	4.2	2.4	16.3	Canada
EU(15)	156.8	150.2	141.0	18.8	24.0	-5.3	135.8	UE(15)
Other Europe	38.9	28.0	29.1	15.1	15.3	-0.2	28.9	Autres Europe
Japan	87.8	84.8	81.7	4.2	19.6	-15.4	66.4	Japon
Australia and New Zealand	9.0	9.0	8.1	1.5	3.2	-1.7	6.4	Australie et Nouvelle-Zélande
Korea	39.3	38.4	36.7	4.7	15.6	-10.9	25.8	Corée
Mexico	15.1	12.5	12.6	1.0	4.2	-3.2	9.5	Mexique
OECD	449.9	426.2	402.6	82.8	87.9	-5.1	397.5	OCDE
Brazil	24.6	19.8	21.6	0.7	8.8	-8.1	13.5	Brésil
OECD Steel Committee	474.5	446.0	424.2	83.5	96.7	-13.2	411.0	Comité de l'acier de l'OCDE
Other Latin America	9.1	8.2	7.9	7.8	4.1	3.7	11.6	Autres Amérique latine
South Africa	6.4	6.3	5.8	0.3	1.8	-1.6	4.2	Afrique du Sud
Other Africa	0.7	0.3	0.6	3.4	0.2	3.2	3.8	Autres Afrique
Middle East	13.3	13.0	12.0	16.1	0.9	15.1	27.1	Moyen-Orient
India	25.2	12.5	21.1	1.5	1.3	0.2	21.2	Inde
ASEAN(5)	8.3	8.3	7.5	14.2	2.8	11.4	18.9	ASEAN(5)
Other Asia	19.2	18.7	17.9	13.5	4.3	9.2	27.1	Autres Asie
Total	82.2	67.3	72.8	56.8	15.4	41.2	113.9	Total
Central and Eastern Europe	10.8	6.4	8.7	1.4	5.6	-4.3	4.4	Europe centrale et orientale
Of which:								dont :
Romania	5.8	2.9	4.6	0.4	2.1	-1.7	2.9	Roumanie
Slovak Republic	3.1	3.1	2.6	0.7	2.4	-1.7	0.9	République slovaque
NIS	76.1	31.2	56.6	4.7	35.5	-30.8	25.8	NEI
Of which:								dont :
Russia	44.7	23.1	33.9	1.6	21.5	-19.9	14.0	Russie
Ukraine	25.1	5.0	18.0	0.6	11.5	-10.9	7.1	Ukraine
Other NIS	6.3	3.0	4.7	2.5	2.5	-0.0	4.7	Autres NEI
China	110.2	77.1	95.2	13.0	4.5	8.5	103.7	Chine
World	754.1	627.9	657.5	159.2	157.8	1.5	659.0	Monde

Source: OECD/OCDE.

52

Table 7. **Steel production, consumption and trade** (million tonnes)
Tableau 7. **Production, consommation et échanges d'acier** (en millions de tonnes)

2000	Production Crude steel/ Acier brut	Production via c.c.	Production Product eq./ Équiv. produits	Imports/ Importations	Exports/ Exportations	Balance/ Solde	Apparent consumption/ Consommation apparente	2000
United States	99.1	93.7	85.5	29.4	4.6	24.8	110.3	États-Unis
Canada	16.1	15.9	13.9	6.9	3.8	3.1	17.0	Canada
EU(15)	161.2	154.4	145.0	18.5	25.5	-7.0	138.0	UE(15)
Other Europe	36.5	30.4	31.3	14.7	16.2	-1.5	29.8	Autres Europe
Japan	89.6	86.6	83.5	4.0	19.2	-15.2	68.3	Japon
Australia and New Zealand	8.3	8.3	7.5	1.6	2.5	-0.9	6.6	Australie et Nouvelle-Zélande
Korea	40.2	39.6	37.5	4.9	12.2	-7.4	30.2	Corée
Mexico	16.0	12.5	13.3	1.2	4.4	-3.2	10.1	Mexique
OECD	466.8	441.3	417.4	81.1	88.4	-7.3	410.2	OCDE
Brazil	25.3	20.4	22.2	0.5	8.6	-8.1	14.1	Brésil
OECD Steel Committee	492.1	461.7	439.6	81.6	97.0	-15.4	424.3	Comité de l'acier de l'OCDE
Other Latin America	10.4	9.1	9.0	7.9	4.5	3.5	12.4	Autres Amérique latine
South Africa	6.7	6.5	6.0	0.3	1.9	-1.6	4.4	Afrique du Sud
Other Africa	0.8	0.4	0.7	3.5	0.2	3.3	4.0	Autres Afrique
Middle East	14.7	14.4	13.2	17.0	1.0	16.0	29.2	Moyen-Orient
India	26.0	13.0	21.8	1.9	1.5	0.5	22.2	Inde
ASEAN(5)	8.9	8.9	8.1	15.3	2.6	12.7	20.7	ASEAN(5)
Other Asia	19.5	18.6	18.1	15.1	3.4	11.7	29.7	Autres Asie
Total	86.4	70.3	76.3	61.0	15.3	46.1	122.6	Total
Central and Eastern Europe	12.4	7.3	10.0	1.4	6.5	-5.1	4.9	Europe centrale et orientale
Of which:								dont :
Romania	6.4	3.2	5.1	0.4	2.4	-2.0	3.1	Roumanie
Slovak Republic	3.5	3.5	3.0	0.7	2.7	-2.0	1.0	République slovaque
NIS	78.2	32.4	58.3	4.5	34.5	-30.0	28.3	NEI
Of which:								dont :
Russia	45.9	23.7	34.8	1.5	21.0	-19.5	15.3	Russie
Ukraine	25.2	5.0	18.1	0.5	10.5	-10.0	8.1	Ukraine
Other NIS	7.2	3.6	5.4	2.5	3.0	-0.5	4.9	Autres NEI
China	115.7	81.0	99.9	11.4	4.5	6.9	106.8	Chine
World	785.8	653.2	684.9	159.9	158.5	1.4	686.2	Monde

Source: OECD/OCDE.

Table 8. The steel markets in the United States, the European Union and Japan
Tableau 8. **Les marchés de l'acier aux États-Unis, dans l'Union Européenne et au Japon**

	United States/États-Unis				EU(15)/UE(15)				Japan/Japon				
	1997	1998	1999	2000	1997	1998	1999	2000	1997	1998	1999	2000	
	In million product tonnes/En millions de tonnes produit												
Real consumption	108.2	114.0	108.3	111.0	131.9	134.6	138.8	141.0	80.2	67.9	66.8	68.1	Consommation réelle
Stocks, consumers and merchants	+0.1	+1.6	-1.0	-0.5	+1.0	+3.0	-2.0	-1.5	+0.8	-0.8	-0.5	+0.2	Stocks des consom. et des marchands
Market	108.3	116.6	107.3	110.5	132.9	137.6	136.8	139.5	81.0	67.1	66.3	68.3	Marché
Imports	28.9	37.7	31.4	29.4	16.4	23.4	18.8	18.5	6.4	4.8	4.2	4.0	Importations
Exports	5.6	5.0	4.4	4.6	28.1	24.0	24.0	25.5	22.8	24.9	19.6	19.2	Exportations
Deliveries	85.0	83.9	80.3	85.7	144.6	138.2	142.0	146.5	97.4	87.2	81.7	83.5	Livraisons
Producers' stocks	+0.1	+1.4	-0.2	-0.2	-1.0	+5.4	-1.0	-1.5	0	0	0	0	Stocks des prod.
Production	85.1	84.3	80.1	85.5	143.6	143.6	141.0	145.0	97.4	87.2	81.7	83.5	Production
	In million tonnes of crude steel/En millions tonnes d'acier brut												
Crude steel production	98.5	97.7	92.8	99.1	159.8	159.6	156.8	161.2	104.6	93.6	87.8	89.6	Production d'acier brut
Capacity	109.3	113.3	116.3	116.3	196.1	198.2	197.8	198.2	149.5	149.5	149.5	149.5	Capacité
	In %/En %												
Capacity utilisation	90.1	85.9	79.8	85.2	81.5	80.7	79.3	81.3	70.0	62.3	58.7	60.0	Utilisation de capacité
Import share	26.7	32.7	29.9	26.6	12.3	17.0	13.7	13.3	8.0	7.2	6.3	5.9	Part d'importation

Source: OECD/OCDE.

54

Table 9. **Steel markets in EU countries** (million tonnes of product equivalent)

Tableau 9. **Les marchés de l'acier dans les pays de l'UE** (en millions de tonnes d'équivalent produits)

Germany/Allemagne

	1997	1998	1999	2000	
Real consumption	37.9	39.3	36.5	36.3	Consommation réelle
Stocks	-0.4	+0.5	-0.1	-0.2	Stocks
Apparent cons.	38.3	39.8	36.4	36.1	Consomm. apparente
Imports	15.5	18.5	16.2	16.4	Importations
Exports	17.7	18.2	18.4	19.1	Exportations
Production	40.4	39.6	38.6	38.8	Production
Crude steel prod.	45.0	44.1	42.9	43.3	Prod. d'acier brut
Capacity	51.6	51.7	51.7	51.7	Capacité
Capacity utilisation in %	87	85	83	84	Utilisation de la capacité en %

France

	1997	1998	1999	2000	
Real consumption	17.0	18.9	19.2	18.3	Consommation réelle
Stocks	-0.5	+0.5	-0.5	-0.2	Stocks
Apparent cons.	16.5	19.4	18.7	18.1	Consomm. apparente
Imports	10.6	14.1	13.9	12.4	Importations
Exports	11.9	12.8	13.1	12.7	Exportations
Production	17.7	18.1	17.9	18.4	Production
Crude steel prod.	19.8	20.1	20.0	20.6	Prod. d'acier brut
Capacity	24.2	24.3	24.3	24.3	Capacité
Capacity utilisation in %	82	83	82	85	Utilisation de la capacité en %

United Kingdom/Royaume-Uni

	1997	1998	1999	2000	
Real consumption	13.5	13.6	14.1	14.8	Consommation réelle
Stocks	+0.4	+0.8	-0.2	-0.3	Stocks
Apparent cons.	13.9	14.4	13.9	14.5	Consomm. apparente
Imports	5.3	6.4	6.0	6.0	Importations
Exports	7.9	7.5	7.2	7.3	Exportations
Production	16.5	15.5	15.1	15.8	Production
Crude steel prod.	18.5	17.3	16.8	17.6	Prod. d'acier brut
Capacity	20.2	20.3	20.3	20.3	Capacité
Capacity utilisation in %	92	85	83	87	Utilisation de la capacité en %

Italy/Italie

	1997	1998	1999	2000	
Real consumption	26.1	25.8	26.5	27.8	Consommation réelle
Stocks	-1.0	+0.5	-1.0	-0.8	Stocks
Apparent cons.	25.1	26.3	25.5	27.0	Consomm. apparente
Imports	10.4	12.1	11.9	12.3	Importations
Exports	8.5	9.0	9.0	9.0	Exportations
Production	23.2	23.2	22.5	23.7	Production
Crude steel prod.	25.8	25.8	25.1	26.4	Prod. d'acier brut
Capacity	35.5	35.6	35.6	35.6	Capacité
Capacity utilisation in %	73	72	71	74	Utilisation de la capacité en %

Netherlands/Pays-Bas

	1997	1998	1999	2000	
Real cons.	4.3	4.2	4.4	4.5	Consommation réelle
Stocks	0.0	+0.2	-0.1	-0.1	Stocks
Apparent cons.	4.3	4.4	4.3	4.4	Consomm. apparente
Imports	4.8	5.3	4.2	4.8	Importations
Exports	6.5	6.6	5.5	6.3	Exportations
Production	6.0	5.8	5.6	5.9	Production
Crude steel prod.	6.6	6.4	6.2	6.5	Prod. d'acier brut
Capacity	6.8	6.8	6.8	6.8	Capacité
Capacity utilisation in %	91	94	91	96	Utilisation de la capacité en %

Belgium and Luxembourg/Belgique et Luxembourg

	1997	1998	1999	2000	
Real cons.	5.3	5.4	5.6	6.0	Consommation réelle
Stocks	-0.2	+0.6	-0.2	-0.2	Stocks
Apparent cons.	5.1	6.0	5.4	5.8	Consomm. apparente
Imports	6.6	7.05	8.0	7.2	Importations
Exports	13.5	13.6	14.6	13.7	Exportations
Production	12.0	12.5	11.9	12.3	Production
Crude steel prod.	13.3	13.9	13.2	13.6	Prod. d'acier brut
Capacity	18.2	18.5	18.6	18.8	Capacité
Capacity utilisation in %	73	75	71	72	Utilisation de la capacité en %

Spain/Espagne

	1997	1998	1999	2000	
Real cons.	12.7	14.4	15.3	15.6	Consommation réelle
Stocks	+0.6	+1.5	+0.1	-0.3	Stocks
Apparent cons.	13.3	14.9	15.3	15.3	Consomm. apparente
Imports	6.6	7.8	7.3	6.0	Importations
Exports	5.6	6.2	5.3	4.7	Exportations
Production	12.3	13.3	13.4	14.0	Production
Crude steel prod.	13.7	14.8	14.9	15.5	Prod. d'acier brut
Capacity	18.7	18.6	18.7	18.8	Capacité
Capacity utilisation in %	73	80	80	82	Utilisation de la capacité en %

Rest EU(15)/Reste UE(15)

	1997	1998	1999	2000	
Real cons.	6.0	6.9	7.3	7.5	Consommation réelle
Stocks	+0.4	+0.9	-0.3	-0.2	Stocks
Apparent cons.	6.4	7.8	7.0	7.3	Consomm. apparente
Imports	5.7.	6.9	5.9	6.3	Importations
Exports	2.0	2.0	1.8	1.9	Exportations
Production	2.8	2.9	2.9	2.9	Production
Crude steel prod.	3.1	3.2	3.2	3.2	Prod. d'acier brut
Capacity	6.1	6.5	6.0	6.0	Capacité
Capacity utilisation in %	51	49	53	53	Utilisation de la capacité en %

Austria/Autriche

	1997	1998	1999	2000	
Real consumption	2.8	3.0	3.2	3.3	Consommation réelle
Stocks	+0.2	+0.4	-0.1	-0.1	Stocks
Apparent cons.	3.0	3.1	3.1	3.2	Consomm. apparente
Imports	2.1	2.1	2.2	2.2	Importations
Exports	3.8	3.8	3.7	3.6	Exportations
Production	4.7	4.7	4.7	4.7	Production
Crude steel prod.	5.2	5.3	5.2	5.3	Prod. d'acier brut
Capacity	5.6	5.6	5.5	5.5	Capacité
Capacity utilisation in %	93	95	95	96	Utilisation de la capacité en %

Finland/Finlande

	1997	1998	1999	2000	
Real cons.	2.5	2.5	2.7	2.9	Consommation réelle
Stocks	-0.1	+0.1	+0.1	-0.1	Stocks
Apparent cons.	2.4	2.6	2.8	2.8	Consomm. apparente
Imports	1.1	1.2	1.0	0.9	Importations
Exports	2.1	2.3	2.0	1.9	Exportations
Production	3.5	3.7	3.8	3.8	Production
Crude steel prod.	3.7	4.0	4.0	4.1	Prod. d'acier brut
Capacity	4.2	4.3	4.3	4.3	Capacité
Capacity utilisation in %	88	93	92	95	Utilisation de la capacité en %

Sweden/Suède

	1997	1998	1999	2000	
Real consumption	3.9	4.1	4.1	4.0	Consommation réelle
Stocks	-0.2	+0.3	-0.5	-0.5	Stocks
Apparent cons.	3.7	4.4	3.6	3.5	Consomm. apparente
Imports	2.9	3.3	2.8	2.6	Importations
Exports	3.8	3.6	4.0	3.8	Exportations
Production	4.6	4.7	4.8	4.7	Production
Crude steel prod.	5.2	5.2	5.3	5.2	Prod. d'acier brut
Capacity	5.8	6.0	6.1	6.3	Capacité
Capacity utilisation in %	90	87	87	83	Utilisation de la capacité en %

Note: Trade figures for individual EU-countries represent the sum of intra-EU trade and trade with third countries.

Note : Les chiffres d'échanges pour les pays individuels des UE représentent la somme des échanges avec les pays tiers et des échanges intra-communautaires.

Source: OECD/OCDE.

Table 10. **Steel markets in "other Western Europe" and in Mexico** (million tonnes of product equivalent)

Tableau 10. **Les marchés de l'acier dans les autres pays d'Europe occidentale et le Mexique** (en millions de tonnes d'équivalent produits)

	Turkey/Turquie				Iceland and ex-Yugoslavia/ Islande et ex-Yougoslavie				Norway/Norvège				Switzerland/Suisse				
	1997	1998	1999	2000	1997	1998	1999	2000	1997	1998	1999	2000	1997	1998	1999	2000	
Apparent consumption	10.8	11.5	11.8	12.6	1.4	1.5	1.4	1.4	1.5	1.6	1.4	1.3	1.9	1.9	1.9	2.0	Consommation apparente
Imports	5.6	5.8	6.0	4.3	1.1	1.1	1.0	1.2	1.7	1.9	1.6	1.5	1.9	2.0	2.0	2.0	Importations
Exports	7.6	7.0	6.6	6.3	1.1	1.0	0.6	1.0	0.8	0.8	0.7	0.7	0.9	1.0	1.0	1.0	Exportations
Production	12.8	12.7	12.4	14.6	1.4	1.4	1.0	1.2	0.5	0.6	0.5	0.5	1.0	0.9	0.9	1.0	Production
Crude steel production	14.2	14.1	13.8	16.3	1.5	1.6	1.1	1.4	0.6	0.6	0.6	0.6	1.1	1.0	1.0	1.1	Production d'acier brut
Capacity	19.5	19.9	19.9	20.1	2.5	2.8	2.8	2.8	0.6	0.6	0.6	0.6	1.1	1.1	1.1	1.1	Capacité
Capacity utilisation in %	73	71	69	81	60	57	39	50	100	100	100	100	100	97	93	100	Utilisation de la capacité en %

	Czech Republic/ République tchèque				Hungary/Hongrie				Poland/Pologne				Mexico/Mexique				
	1997	1998	1999	2000	1997	1998	1999	2000	1997	1998	1999	2000	1997	1998	1999	2000	
Apparent consumption	3.8	4.1	3.6	3.9	1.4	1.7	1.8	1.9	6.5	6.8	7.1	7.4	7.6	9.2	9.5	10.1	Consommation apparente
Imports	2.2	2.5	2.3	2.0	0.9	1.1	1.2	1.2	1.4	1.9	1.8	1.4	1.6	2.2	1.0	1.2	Importations
Exports	3.9	3.7	3.4	3.7	1.0	1.0	1.0	0.9	4.2	3.2	2.6	2.6	6.0	5.0	4.2	4.4	Exportations
Production	5.5	5.3	4.7	5.5	1.4	1.6	1.6	1.6	9.2	8.0	7.9	8.6	12.0	11.9	12.6	13.3	Production
Crude steel production	6.8	6.5	5.7	6.7	1.7	1.8	1.9	1.9	11.6	9.9	9.7	10.6	14.3	14.2	15.1	16.0	Production d'acier brut
Capacity	8.8	8.8	8.8	8.8	1.9	1.9	1.9	1.9	12.5	12.3	12.4	12.4	15.6	18.1	18.5	18.9	Capacité
Capacity utilisation in %	77	74	65	76	92	95	100	100	93	80	78	85	92	78	82	85	Utilisation de la capacité en %

Source: OECD/OCDE.

Table 11. **Manpower**
Tableau 11. **Main-d'œuvre**

	Average numbers employed ('000) / Moyenne des effectifs ('000)							% change / Variation en %	
	1974	1984	1994	1995	1996	1997	1998	1998/97	1998/74
Belgium/Luxembourg	86.6	51.4	30.9	29.8	28.4	25.3	25.2	-0.4	-70.9
Denmark, Ireland	3.5	2.3	1.6	1.5	1.5	1.6	1.6	0	-54.3
France	155.7	87.1	40.4	39.3	38.5	38.3	38.3	0	-75.4
Germany	230.6	156.5	100.0	92.5	85.9	82.3	80.2	-2.6	-65.2
Greece	8.7	4.2	2.7	2.5	2.3	2.1	2.0	-4.8	-77.0
Italy	93.8	81.7	45.5	42.1	39.2	37.0	38.5	+4.1	-59.0
Netherlands	23.8	18.7	13.1	12.6	12.3	12.1	11.9	-1.7	-50.0
Portugal	5.0	6.7	2.9	2.7	2.4	2.1	1.9	-9.5	-62.0
Spain	89.4	69.2	26.7	25.3	23.7	23.2	23.0	-0.9	-74.3
United Kingdom	197.7	62.3	38.5	37.9	37.0	35.9	33.8	-5.8	-82.9
EU(12)	894.8	540.1	302.3	286.2	271.2	259.5	256.4	-1.2	-71.3
Austria	43.0	34.9	15.4	13.3	12.9	12.4	12.2	-1.6	-71.6
Finland	8.1	9.0	8.8	7.2	7.4	7.3	7.2*	-1.4	-71.1
Sweden	51.0	32.2	20.7	14.5	14.0	13.8	13.4	-2.9	-73.7
EU(15)	996.9	616.2	347.2	321.2	305.6	293.3	289.2*	-1.4	-71.0
Norway	7.3*	4.0	1.3	1.3*	1.3*	1.3*	1.0	-20.0	-85.8
Switzerland	5.2	3.0	1.6	1.3	1.2	1.0	0.7	-30.0	-86.5
Turkey	36.1	35.0	32.4	29.9*	31.0	29.9	29.9*	0	-17.2
Canada	52.2	51.5	31.5	33.7	33.5	33.4	34.5	+3.3	-33.9
United States	609.5	267.4	233.5	240.7	237.6	236.0	232.8	-1.4	-61.8
Australia	43.2	30.5	26.0*	26.0*	26.0*	24.0*	24.0	0	-44.4
Japan	323.9	264.8	182.7	168.8	155.1	146.4	137.9	-5.8	-57.4
Mexico	45.5	77.6*	57.0	60.0	60.0	60.0	60.0*	0	+31.9
Korea	62.9*	62.9	59.8*	66.3*	66.5	65.8	59.5	-9.6	-5.4
Total OECD	2 183.0*	1 412.9*	1 013.4	937.8*	913.7*	887.1*	869.5	-2.0	-60.2
Czech Republic	..	..		78.6		12.9..	..		
Hungary	..	..	16.9	16.0	14.6		..		
Poland	..	..	95.9	95.5	91.6	89.0	..		

Belgique/Luxembourg
Danemark, Irlande
France
Allemagne
Grèce
Italie
Pays-Bas
Portugal
Espagne
Royaume-Uni
UE(12)
Autriche
Finlande
Suède
UE(15)
Norvège
Suisse
Turquie
Canada
États-Unis
Australie
Japon
Mexique
Corée
Total OCDE
République tchèque
Hongrie
Pologne

* Secretariat estimate.   * Estimation du Secrétariat de l'OCDE.
Source: OECD/OCDE.

OECD PUBLICATIONS, 2, rue André-Pascal, 75775 PARIS CEDEX 16
PRINTED IN FRANCE
(58 1999 03 1 P) ISBN 92-64-17139-8 – No. 50955 1999